Thomas Wharton Jones

Failure of Sight from Railway and Other Injuries of the Spine and

Head

Its nature and treatment, with a physiological and pathological disquisition into the

influence of the vaso-motor nerves on the circulation of the blood in the extreme

vessels

Thomas Wharton Jones

Failure of Sight from Railway and Other Injuries of the Spine and Head
*Its nature and treatment, with a physiological and pathological disquisition into the influence
of the vaso-motor nerves on the circulation of the blood in the extreme vessels*

ISBN/EAN: 9783337393755

Printed in Europe, USA, Canada, Australia, Japan

Cover: Foto ©berggeist007 / pixelio.de

More available books at **www.hansebooks.com**

FAILURE OF SIGHT

FROM

RAILWAY AND OTHER INJURIES

OF THE

SPINE AND HEAD

ITS

NATURE AND TREATMENT

WITH A PHYSIOLOGICAL AND PATHOLOGICAL DISQUISITION INTO
THE INFLUENCE OF THE VASO-MOTOR NERVES ON THE
CIRCULATION OF THE BLOOD IN THE
EXTREME VESSELS

BY

THOMAS WHARTON JONES, F.R.S., F.R.C.S.

PROFESSOR OF OPHTHALMIC MEDICINE AND SURGERY IN UNIVERSITY COLLEGE,
LONDON; OPHTHALMIC SURGEON TO THE HOSPITAL; CORRESPONDING
MEMBER OF THE IMPERIAL-ROYAL SOCIETY OF PHYSICIANS OF
VIENNA; MEMBER OF THE ROYAL MEDICAL SOCIETY
OF COPENHAGEN; MEMBER OF THE SOCIETY
OF BIOLOGY OF PARIS, ETC., ETC.

LONDON
JAMES WALTON
BOOKSELLER AND PUBLISHER TO UNIVERSITY COLLEGE
137 GOWER STREET
1869.

PREFACE.

"Nil fingendum, nil excogitandum, sed inveniendum quod Natura ferat, quod Natura faciat."—*Bacon.*

THE failure of sight which so often supervenes in cases of concussion of the spinal cord or brain is amaurotic in its character, and appears to depend immediately on a disturbance of the circulation of the blood in the optic nervous apparatus.

How disturbance of the circulation in the optic nervous apparatus should be induced by concussion of the brain, seems intelligible enough; but how such an effect should supervene on injury of the spinal cord, is a question which is not so easily answered at first.

A consideration of the source, in the spinal cord, of the nervous influence by which the circulation of the blood in the optic nervous apparatus is regu-

a 2

lated, appearing to me calculated to lead to a solution of the question, I have made a physiological and pathological disquisition into the action which the vaso-motor nerves exert on the constrictions and dilatations of the arteries.

In this disquisition I have necessarily entered, with some detail, into the subject of hyperæmia, or vascular congestion, in its various degrees, from simple determination of blood to actual inflammatory stasis.

While, therefore, the work which I now publish is a practical Treatise on the Nature and Treatment of Failure of Sight from Injury of the Spine and Head, it is offered also as a Treatise on the Physiology and Pathology of the Circulation of the Blood in the Extreme Vessels—a subject which, on account of its fundamental importance in the Science of Medicine and Surgery, claims from every Medical man the most careful study.

To turn now to the medico-legal aspect of the question. The present work was originally undertaken for the purpose of aiding in the diagnosis of the existence or non-existence of amaurotic failure of sight in cases of railway injury, in which there was a conflict of opinion on the subject. The observations founded on the cases, and the inferences deduced from them, are such as suggested themselves at the time of drawing up reports for the guidance of those engaged in the legal inquiry.

Throughout the work, therefore, the medico-legal bearings of the subject have been constantly kept in view.

T. WHARTON JONES.

35, GEORGE STREET, HANOVER SQUARE,
London, September 15th, 1869.

CONTENTS.

—◆—

CHAPTER IV.

CHAPTER V.

CHAPTER VI.

CHAPTER VII.

CHAPTER VIII.

CHAPTER IX.

CHAPTER X.

CHAPTER XI.

CHAPTER XII.

CHAPTER XIII.

CHAPTER XIV.

CHAPTER XV.

CHAPTER XVI.

CHAPTER XVII.

CHAPTER XVIII.

CHAPTER XIX.

PART SECOND.

PATHOLOGY OF THE AMAUROTIC FAILURE OF SIGHT CAUSED BY RAILWAY AND OTHER INJURIES OF THE BRAIN.

CHAPTER I.

CHAPTER II.

CHAPTER III.

CHAPTER IV.

PART THIRD.

DIAGNOSIS, PROGNOSIS, AND TREATMENT IN CASES
OF AMAUROTIC FAILURE OF SIGHT FROM RAIL-
WAY AND OTHER INJURIES OF THE SPINE AND
HEAD.

CHAPTER I.

CHAPTER II.

CHAPTER III.

CHAPTER IV.

APPENDIX:

COMPRISING AN ADDITIONAL CHAPTER ON INFLAMMATION.

ADDITIONAL CHAPTER.

FAILURE OF SIGHT

FROM

RAILWAY AND OTHER INJURIES

OF THE

SPINE AND HEAD:

ITS

NATURE AND TREATMENT.

GENERAL ARGUMENT.

THE pathology of the amaurotic failure of sight caused by injuries of the spinal cord is considered in PART FIRST; and the pathology of the amaurotic failure of sight caused by injuries of the brain, in PART SECOND.

The injury sustained being sometimes of a mixed spinal and cerebral character, the sight may be found affected primarily through the brain, as well as secondarily through the spinal cord. The nature of such cases appears from what is said under the proper heads.

PART THIRD comprises the diagnosis, prognosis, and treatment in cases of amaurotic failure of sight from injuries of the spine and head.

In certain idiopathic or spontaneous diseases of the spinal cord and brain, failure of sight is also found to supervene. Such cases are noticed here and there, for the purpose of illustrating the nature and treatment of the traumatic affections, which more especially con- stitute the subject of the following pages.

In an APPENDIX an additional chapter on Inflam- mation is given, in which certain points relating to the process, which it would have been out of place to have entered on in the text, are noticed on account of their general pathological importance.

PART FIRST.

PATHOLOGY OF THE AMAUROTIC FAILURE OF
SIGHT CAUSED BY RAILWAY AND OTHER
INJURIES OF THE SPINAL CORD.

B

CHAPTER I.

INTRODUCTORY OBSERVATIONS.

In cases of concussion of the spinal cord, an affection of the sight, in the form of perverted, impaired, or lost sensibility of the optic nervous apparatus, is liable to supervene. The symptoms come on for the most part insidiously, and do not, until some considerable time after the accident, attain any great degree of prominence.

Railway collisions have of late been a frequent cause of such injuries.

Sometimes the affection of the sight has been the damage most complained of. Sometimes, though not less serious in degree, it has been comparatively secondary in importance, owing to the very grave nature of the other injuries suffered by the unfortunate victim of the accident. Broken down in body and mind—his muscular power gone, his memory lost, and his thoughts confused—the patient is unable to look after his affairs as before ; he cannot fix his attention on any subject, and the concentration of mind in concert or co-operation with the eyes and hands, so necessary in various employments, has become impossible to him. The

B 2

sense of touch has been found actually so obtuse, that the patient could not feel the difference between the pile of cloth and that of velvet, nor distinguish a sovereign from a shilling by the difference in weight. His hearing, also, is sometimes impaired, and *tinnitus aurium* adds to the confusion and distraction of his thoughts.

Mr. Erichsen, in the able and instructive lectures on Railway Injuries, which he delivered at University College Hospital some three years ago, and which he has since published,* called attention to a remarkable feature in the history of such cases, viz., the disproportion between the apparently trifling injury and the really serious mischief that so often occurs in consequence of it. There may, in fact, have been little more than a violent jolting of the body backwards and forwards, without any actual blow on the back ; and the person may, on recovering from the alarm occasioned by the collision, have even supposed himself uninjured. After a time, however, he feels out of sorts, and unfit for work.

Symptoms of spinal injury now slowly develope themselves. In the early stages, says Mr. Erichsen, the chief complaint is a sensation of lassitude, weariness, and inability for mental and physical exertion. Then come the pains, tinglings, and numbness of the limbs ; next the fixed pain and rigidity of the spine ; then the mental confusion and signs of cerebral dis-

* "On Railway and other Injuries of the Nervous System," London, 1866.

turbance, and the affection of the organs of sense; the loss of motor power, and the peculiarity of gait.

The period of the supervention of these symptoms after the occurrence of the injury varies greatly. Most commonly, after the first and immediate effects of the accident have passed off, there is a period of comparative ease and of remission of the symptoms, during which the patient imagines that he will speedily regain his health and strength. This period may last for many weeks, possibly for two or three months. At this time there will be considerable fluctuation in the patient's state. So long as he is at rest he will feel tolerably well; but any attempt at ordinary exertion of body or mind brings back all the feelings or indications of nervous prostration and irritation so characteristic of the injuries under notice; and to these symptoms will gradually be superadded the more serious ones which evidently proceed from a chronic disease of the cord and its membranes. After a lapse of several months—from three to six—the patient will find that he is slowly but steadily becoming worse, and he then, perhaps, for the first time becomes aware of the serious and deep-seated injury that his nervous system has sustained.

As regards the eyes, they may not in appearance present much the matter with them. The symptoms are at first chiefly subjective. The patient may even still be able to make out the smallest print, but he cannot continue to read it longer than a few minutes at a time. Perhaps, however, it is with one eye only that he can do this much; the sight of the other being found, on trying it alone, much impaired. Perhaps the sight of both eyes is dim. The dimness may be in the form of a haze coming on for a time only, and going off again, or it may be persistent.

The patient very often, though not always, complains of seeing luminous spectra, in the shape of rings, stars, sparks, and flashes of fire before his eyes. He is sometimes also troubled with scotomata or muscæ volitantes.

Sometimes there is intolerance of light, sometimes not.

Frequently the adjusting power of the eyes is impaired.

Sometimes the patient sees double.

That an affection of the sight, such as has been sketched, is a not uncommon occurrence, sooner or later, in cases of concussion of the spinal cord is now a notorious fact. In those cases, however, in which the injury to the spine has been sustained in a railway collision, the reality of the affection of the sight complained of necessarily becomes a question when compensation from the railway company is claimed by the injured person for the damage he has suffered. The reality of the injury to the sight comes to be a question, for the reason especially, that the eyes may not, as has been stated, appear at first to have very much the matter with them, and that the patient may still be able to make out the smallest print. Even under the ophthalmoscope morbid changes of structure may not at first be appreciable, the intracranial part of the optic nervous apparatus being alone as yet affected. Sooner or later, however, morbid appearances in the eye do become evident.

CHAPTER II.

FROM Mr. Erichsen's lectures I will quote, in the first place, the particulars relating to the failure of the sight, in four of the cases he has reported.

CASE I.

Mr. R., æt. 35, was in November, 1864, much and severely shaken in a railway collision, but did not suffer any external injury, except a blow upon his face and a cut of his upper lip on the left side. In February, 1866, fifteen months after the accident, Mr. Erichsen saw the patient, and found him pallid in the face, with an expression of habitual suffering, and unable to stand or walk without support.

Since the accident Mr. R.'s memory has been bad. He cannot add up an ordinary sum, and is now unable to transact any business ; can bear neither light nor noise. He complains of stars, sparks, flashes of light and coloured spectra flaming and flashing before his eyes. He cannot read for more than two or three minutes at a time, the letters appearing confused, and the effort being painful to bear.

On examining the state of the eyes Mr. Erichsen found that vision was good in the right eye, but that this organ was over-sensitive to light. In the left eye vision was nearly lost, so much so that Mr. R. could not read large print with it.

The hearing of the right ear was over-sensitive ;— that of the left, dull.

Mr. R. complains of a numbness, with tingling and burning sensations in the right arm and leg, but makes no complaint of the left side. These sensations are worst in the morning.

Even while supporting himself, if he attempts to rest on his right foot, the leg sinks, as it were, under him.

Mr. Erichsen found the right leg objectively colder than the left. To the patient's feeling both legs and feet were cold.

The patient kept the body perfectly straight, fixed, and immoveable, as he could not bend the spine in any direction without suffering severe pain.

There was considerable pain at the occipito-atlantal articulation, as also at that between the atlas and dentata. If an attempt was made to bend the head forcibly forward, or to rotate it, such severe pain was occasioned, that it became necessary to desist. When directed to look round, the patient turned the whole body.

On pressure, tender spots were found in the upper cervical, the middle dorsal and lumbo-sacral regions.

Since the accident, it has been remarked that Mr. R.

has become unable to judge correctly of the distance of objects in a *lateral* direction, though he appears to be able to do so when looking straight forward. Thus when driving in the middle of a straight road, he always imagines that the carriage is in danger of running into the ditch or hedge on the *near* side.

The opinion which Mr. Erichsen formed of the nature of the case was, that the patient had sustained an injury of the spinal cord, and that the base of the brain was also to some extent, though probably secondarily, involved, and that chronic sub-acute meningitis of the spine and base of the cranium had taken place. In accordance with this view, his prognosis was altogether unfavourable.

CASE II.

Mr. I., æt. 43, was suddenly dashed forwards, and then violently thrown backwards, in a railway collision on the 23rd of August, 1864. He did not experience any immediate suffering from the shock, but on his return home the same evening, he began to feel ill.

On March 8, 1865, Mr. Erichsen saw the patient in consultation, and found him labouring under the following, among other, subjective symptoms :—Loss of memory; confusion of thought and ideas ; disturbed sleep; pains and noises in the head ; partial deafness of the left ear, morbid sensibility of the right ; irritability of the eyes, rendering light very painful, though vision had become imperfect in the right eye ; numbness, tingling sensation and formication in the right arm and leg.

Mr. I. walked with a peculiar, unsteady, straddling gait, and had to use a stick or otherwise to support himself. He could stand for a moment on the left leg, but immediately fell over if he attempted to do so on the right.

His right arm and hand were numb; the little and ring fingers contracted.

The spine was tender and painful in the upper cervical, in the middle dorsal, and in the lower lumbar, regions. Movement of any kind greatly increased the pain. When the head was moved on the atlas, or the atlas on the dentata, agonising pain was occasioned. The spine was therefore kept erect and rigid.

Concussion of the spine causing chronic inflammation of its membranes and of the cord was the diagnosis.

The unfavourable prognosis which Mr. Erichsen gave was justified by the fact that Mr. I., one year and nine months after the accident, remained a broken-down invalid, and unfit for business of any kind.

CASE III.

Captain N., æt. 38, was, in 1854, thrown from a pony-chaise and was much bruised and shaken.

About six months after the accident the following symptoms began to manifest themselves :—Confusion of thought; impaired memory; giddiness, especially on moving the head suddenly; impaired sight, with muscæ volitantes, sparks and flashes of light before his eyes;

inability to read beyond a few minutes, partly because the letters seemed to run into each other, and partly because he could not concentrate his thoughts so as to fix his attention.

He now began to suffer from a feeling of numbness and a sensation of " pins and needles " in both hands, but more particularly the left, and chiefly in those parts supplied by the ulnar nerve. He complained of the same sensations in the left leg and foot. He walked with difficulty, and with the legs somewhat apart, using a stick, or else supporting himself by holding on to pieces of furniture in the room as he passed them. He can stand on the right leg, but the left one immediately gives way under him.

The spine is tender on pressure and on percussion in the lower cervical region and between the shoulders. It is stiff; he cannot bend the back without pain, and cannot stoop without falling forward.

When Captain N. consulted Mr. Erichsen on the 27th of October, 1862, i.e., eight years after the accident, he was still labouring under the same train of symptoms, though not in quite so severe a degree as before.

CASE IV.

Mr. C. W. E., æt. 50, was violently shaken to and fro in a railway collision on the 3rd of February, 1865.

Fourteen months after the accident, Mr. Erichsen found him affected with loss of memory and confusion of thoughts, and unable to concentrate his attention beyond a few minutes upon any one subject. If he attempts to read, he is obliged to lay aside the paper or book in a few minutes, as the letters become blurred and confused. If he tries to write, he often misspells the commonest words, but he has no difficulty about figures. He is troubled with horrible dreams, and wakes up frightened and confused. His head is habitually hot, and often flushed. He complains of a dull confused sensation within it, and of loud noises, which are constant.

The hearing of the right ear is dull. He cannot hear the tick of an ordinary watch at a distance of six inches from it; with the left ear he can hear the tick at twenty inches distance.

The vision of the left eye has been weak from childhood; that of the right, which was previously good, has become seriously impaired since the accident. The patient is troubled with muscæ volitantes, and sees a fixed line or bar, vertical in direction, in the field of vision. He complains also of the appearance of flashes, stars, and coloured rings before his eyes. Light—even ordinary daylight—is especially distressing to him. In fact, the eye is so irritable that he has an abhorrence of light. He habitually sits in a darkened room, and cannot bear to look at artificial light, such as that of gas, candles, or the fire.

He often thinks that he smells fetid odours not appreciable to others, and has lost his sense of taste to a great degree.

He complains of some numbness, and of " pins and needles " in the left arm and leg, also of pains in the left leg, and a feeling of tightness or constriction. All these symptoms are worst on first rising in the morning.

He walks with great difficulty without support. He can stand on the right leg, but if he attempts to do so on the left, it immediately bends and gives way under him.

The spine is tender on pressure and on percussion at the lower cervical, the middle dorsal, and lumbar regions. The pain is increased on moving the body in any direction, but especially the antero-posterior.

Mr. Erichsen's diagnosis was—that in consequence of the concussion of the spine, inflammatory action of a chronic character had been set up in the Meninges of the Cord, that the partial paralysis of the left leg was probably dependent on disease of the cord itself, and that the cerebral symptoms indicated implication of the brain and its membranes.

The next cases to be adduced are selected from among those which have come under my own observation, and in which I had the opportunity of examining the condition of the eyes in detail.

CASE V.

Mr. W. T., æt. 37, was in a train which ran off the line, but was not conscious of having suffered any injury at the time. Next day, however, he fainted twice while dressing, and excruciating pain in the eyes came on. In consequence of this, Mr. T. made no attempt to exert the sight for a fortnight, and then on trying to read, found that, though he saw distinctly enough at first, he could not continue the exertion of the sight any length of time before the words appeared dim and confused, and pain over the eyes and in the temples came on.

Another symptom was *photopsy*, in the form of a dying flame suddenly flaring up.

The sight, before the accident, was good for both near and distant objects, and the patient used to be able to read for any length of time without·fatigue or pain. The sight of the left eye has suffered more than that of the right.

I saw Mr. W. T. three months after the accident. The eyes looked congested. The pupils, on sudden exposure to the light, contracted, but they tended to fall back again to the middle state, especially the pupil of the left eye.

Under the ophthalmoscope, the pupil of the left eye remained wider than that of the right, so that the examination could be better made in the former than in the latter. It was not considered advisable to bring the eyes under the influence of atropia. In the *left* eye the optic disc was reddish, from hyperæmia or vascular injection, and the retinal veins were large, gorged, and dark-looking. In the *right* eye, the appearances, so far as they could be observed through the less dilated pupil, were much the same as in the left.

CASE VI.

R. S., a man æt. about 35, was a patient in University College Hospital under the care of Mr. Erichsen, in the early part of 1868.

Some two years before, he fell from the top of a house on his back and injured his spine. Since then, he has suffered from muscular weakness, especially of the right extremities, and is altogether in a broken-down state of body. His mind and spirits are at the same time much depressed.

Three weeks before my attention was directed to his case, the man found that the sight of his right eye had become so dim that he was scarcely able to make out the largest letters. This failure of the sight was accompanied by luminous appearances, which he compared to fireworks, before his eye, by haloes all round the gas-lights, and by pains in the region of the right eye and the side of the face, extending to the back of the neck.

The left eye presented indications of having been at some former period affected with iritis. On inquiry, the patient stated that the attack of inflammation occurred many years before the accident. The sight of the eye in question, according to the patient's account, did not then suffer, but subsequently to the failure of the sight of the right eye, the sight of the left began to fail also, and that in a similar manner.

Some two or three months before his attention was drawn to the dimness of his right eye, the man found that after reading for five minutes or so, the eyes began to water, and that after reading for half an hour, he had to give up altogether in consequence of an attack of pain over the eyebrow.

The pupil of the right eye contracted freely on sudden exposure to the light.

Under the ophthalmoscope, I observed in the right eye a bluish whiteness of the optic disc on the side next the temple (apparently on the nasal side), with congestion and blackish discoloration of the retina all round, from granular pigmentous deposit—appearances indicative of degeneration of structure.

CASE VII.

Mr. P. O., æt. 48, suffered a shock of the spinal cord in a railway collision. Since then his energies, both bodily and mental, have become much impaired. He is unable to read longer than five or ten minutes at a time without pain in the head coming on, and the words appearing confused and mixed up together.

One year after the injury, dimness of the sight of the left eye was discovered; so that it is now (eighteen months after the injury) with the right eye only that he is able to read at all, even for the short space of time mentioned. Before the accident Mr. O. could see well, both far and near, and could continue to read for hours together.

On examination of the case, eighteen months after the injury, I found that the patient could read print of any size with the right eye, aided by a convex glass fourteen inches focus, but only for five or ten minutes at a time. With the left eye he could see no better than to make out very large letters.

The pupils were natural in their movements.

An ophthalmoscopical exploration gave the following results :—In the *left* eye the optic disc was red from vascular injection—all except the centre, where the retinal trunks emerge, which was unnaturally opaque-white—and, intermixed with the redness, discoloration of a black leaden hue, from granular pigmentous deposit. The retina all round the disc presented similar discoloration. In the *right* eye, there were observed, also, congestion and black discoloration of the disc and retina, though in a less advanced stage.

Mr. O. did not complain of photopsy.

CASE VIII.

Mr. N. L., æt. 34, suffered a shock in a railway collision, and was laid up for three months in consequence. Since then he has found himself unhinged, bodily and mentally.

Ten months subsequently to the injury he first consulted me respecting his eyes, complaining of inability to exert the sight as formerly, and of pain at the bottom of the orbit, extending to the back of the head when he makes the attempt,—of motes and blacks floating before him, and the appearance occasionally as if of flashes of light,—of halos around the lights, and of undue retention of visual sensations. He subsequently complained of a haze which now and then comes over the sight, especially that of the right eye.

In consequence of failure in the power to concentrate his mind and this impaired energy of sight, the patient commits mistakes in summing up and making entries.

Mr. N. L. also complained of an unnatural warm sensation over the head, and other symptoms of perversion of common sensation.

The right eye could not be opened so freely as the left, and it was weak and watery looking. Besides enlarged rectal veins, indicating choroidal congestion, there was some vascular injection of the white of the eye on the temporal side of the cornea. The pupils were active, perhaps rather too much so, indicating an irritable state of the retina.

Under the ophthalmoscope a general fulness of the vessels of the fundus was seen, especially in the right eye. Otherwise, there did not appear to be anything unnatural. Some seven months after the examination just described, there were the same symptoms with more marked disturbance of the circulation in the eyes. The optic discs were the seat of capillary injection, and presented in their temporal halves glimmerings of a bluish whiteness.

In the several examinations of the eyes made subsequently, at intervals of some months, while the subjective symptoms continued much the same, more and more evident indications of degeneration of structure at the fundus were detected, but always in a much less degree in the left eye. The following is the result of the last examination of the right eye, which was instituted just two years and three months from the date of the accident :—The conjunctiva of the lower eyelid, anæmic-looking in the middle, but congested at the edges. The white of the eye, especially next the temporal side of the cornea, still somewhat pink from congestion. Both pupils active, but when the left eye was kept closed, that of the right eye fell back into the middle state of dilatation. Under the ophthalmoscope, the optic disc was observed to be somewhat shrunk and opaque at the margin, especially on the temporal side where it appeared as if retracted from the choroid. It was still reddish from congestion, and presented the bluish white glance in its temporal half before mentioned. The adjacent part of the retina was the seat of increased capillary injection, with slight blackish discoloration.

In this case the congestion was very manifestly of an anæmic character.

CASE IX.

Mr. J. K., æt. 29, was involved in a railway collision, and three or four days thereafter, found that he could not walk except in a tottering manner. About four months after the accident dimness of sight, worse in the right eye, with diplopy, came on. He also complained of dulness of hearing.

The pupils were found to contract on sudden exposure to the light, but immediately to dilate again. The right pupil remained rather wider than the left. After the patient has looked upon an object for some time, the pupils become dilated, but directly he takes a new field they contract again.

One year after the accident I saw this patient, and observed under the ophthalmoscope great congestion of the optic discs. The pupils were wide, and remained so during the examination.

There was some degree of inflammation of the membrana tympani observed.

CHAPTER III.

Loss of memory, confusion of thought, inability to concentrate the attention on any subject, and horrible dreams, were symptoms of mental disturbance, while the perversion or failure of the senses of sight, hearing, smell, taste, and, in part, touch also, in the various modes and degrees mentioned, were symptoms of sensorial disturbance ; both sets of symptoms indicating implication of the brain, either directly at the time of the accident, or consecutively by transmission to the brain of morbid action from the primarily injured spinal cord.

The disturbances of the motor power were chiefly due to the injury of the spinal cord, as we shall see ; but some of them, also, such as the general impairment of the muscular energy, no doubt arose from the cerebral implication.

ASTHENOPIC SYMPTOMS.

Inability of the patient to exert his sight longer than a few minutes, even though he may still be able to make out the smallest print, was, we have seen, a striking symptom in all the cases related. This *as-thenopia*, it is to be observed, is not owing merely to impaired power of maintaining the adjustment of the eyes for the exercise of the sight on near objects, as in the common form of the complaint, but is owing also in part to the erethism or state of irritable weakness in which the eyes are, and which is a manifestation of impaired vital energy of the optic nervous apparatus occasioned by disturbance of the circulation in it.

The patients have sometimes, indeed, appeared to be helped by convex glasses, as in common asthenopia; but it has always struck me that it was as much or more by their mere magnifying power, as in amblyopia or defective sensibility of the retina, that the glasses assisted the sight. The attempt to use the eyes, even when assisted by glasses, often brought on pain in the head, extending from the bottom of the orbit to the occiput, a symptom which belongs to deep-seated disturbance of the circulation in the optic nervous apparatus.

In some cases, though the sight of both eyes was impaired, the patient could see better when one eye was closed. This was owing to a slight degree of diplopia or double vision.

PHOTOPHOBIA, OR INTOLERANCE OF LIGHT.

When the eye is in a morbidly sensitive state, the impression of the external light, its own natural excitant, gives rise to a painfully dazzling glare, which excites, by reflex action, spasmodic closure of the eye-lids; while the patient, still further to relieve his distress, covers his eyes with his hands, and seeks the dark. This constitutes *photophobia*, or intolerance of light.

Intolerance of light is a very common symptom of inflammation of the eye. In the cases related it appears to have been owing to the same state of the optic nervous apparatus as that on which the asthe-nopia depended.

Over sensitiveness of the ear to sounds is analogous in respect both to its nature and cause to intolerance of light. Hyperæsthesia, or over-sensitiveness of the skin to external impressions, is also a state analogous to photophobia, and is often a symptom of irritation of the nervous centre at the roots of the sensitive nerves affected. Notwithstanding the over-sensitiveness, the skin may be, at the same time, numb; just as, along with photophobia, there may be impaired sight.

In some of the cases related the patient suffered greatly from intolerance of light; in other cases, little or not at all.

PHOTOPSIA, OR LUMINOUS SPECTRA.

When a nerve endowed with special sensibility is irritated in any way, the sensation produced is not pain, but some modification of that which is peculiar to the organ of sense, as induced by its own natural excitant. Thus, if the retina or other part of the optic nervous apparatus be irritated, by a disturbed state of the circulation of the blood within it, for example, the sensation of light is called forth, and no other. Such a sensation is named subjective in contradistinction to that sensation of light which is ordinarily excited by the impression of the external agent, *light*, on the retina.

In like manner, the sensation, known by the name of " pins and needles," in the skin is an example of a subjective or referred tactile sensation, called forth independently of any external impression on the skin. Disturbance of the circulation in the nerves of touch, or their roots in the spinal cord, is a common cause of formication, pricking, neuralgia, and other referred tactile sensations, just as a disturbance of the circulation in the optic nervous apparatus may be of the sensation of light.

Anæsthesia, or failure of the sensibility, is, in either case, a common result.

The pain, which may coexist with the special sensation in an irritated organ of sense, has not its seat in the special nervous structure, but is owing to accompanying irritation of some adjacent part which is endowed with common sensibility. This irritation may be of a mechanical nature, or it may be communicated from the primarily irritated nerve of special sensation through the brain, to the nerves of common sensation by the peculiar reflex action known under the name of irradiation of sensations. Thus the pain, extending to the head when the eyes are exposed to very bright light, or when strong impressions are otherwise made on the retina, is indirectly excited; being the result of *irradiation* through the brain from the optic to the fifth nerve.

The subjective sensation of light constitutes what is called Photopsia, or luminous spectra. Photopsia, though not constant, was a very common symptom in the cases related. When present, it is always a very important indication that mischief is going on somewhere in the optic nervous apparatus. But in order fully to understand the significance of photopsia, the different circumstances attending its occurrence must be taken into consideration.

A familiar example of a luminous spectrum occurring independently of external light, though not exactly a *subjective* sensation, is that which, on pressing the eyeball, is seen projected outwards, and on the side opposite to that where the pressure is applied. A blow on the eye excites the appearance of light more vividly. But similar appearances would be seen if the cerebral end of the optic nerve were irritated. Indeed, it is indifferent what part of the optic nervous apparatus be excited, in order that luminous sensations may be perceived,—whether the retina itself be irritated— the fibres of the optic nerve irritated—or whether the cerebral part of the optic nervous apparatus be irritated. When the cause of irritation is from within, the resulting sensation of light is then properly named *subjective*. Moreover, in whatever part the optic nervous apparatus be excited or irritated, the luminous sensation which results, like other visual sensations,

is always referred by the mind to without the body—
i.e., projected outwards in space.

Subjective tactile sensations, like all other sensations of touch, it is to be remembered, are referred by
the mind to the surface of the skin. There may be at
the same time insensibility to external impressions.

Thus it is that appearances of fiery scintillations,
flashes of light, and coloured coruscations before the
eyes, occurring independently of any external impression, are symptoms of irritation of some part of the
optic nervous apparatus—cerebral or ocular—from fulness of blood or the opposite condition of anæmia, or
from other causes. As such a state may end in amaurosis, so the photopsia or luminous spectra are viewed
as symptoms of incipient amaurosis. They may continue to appear, however, after visual sensibility is
nearly or altogether lost.

Photopsy, or subjective sensations of light, though the
sight has failed, is analogous to ringing noises in the
ears, though the hearing is much impaired, and to the
subjective tactile sensations of formication, tingling,
pricking, &c., though the part of the skin to which
they are referred be numb.

UNDUE RETENTION OF THE IMAGES OF SIGHT AND COMPLEMENTARY COLOURS.

The undue retention of the images of sight and
complementary colours were among the symptoms particularly mentioned in some cases.

In the natural state, the images of sight continue to

be seen a short space of time after the impression on the retina which occasioned them has ceased to act. Hence the appearance of an object may continue for a second or two before the eyes, though they have been turned away from looking at it. In certain morbid states of the optic nervous apparatus, spectra from the retention of visual sensations remain after the impression a much longer time than natural, even although vision be otherwise much impaired.

Undue retention of the images of sight is thus a symptom of impaired energy of the optic nervous apparatus, and is a frequent attendant on amaurotic affections.

Analogous retention of sensations in the case of the other senses occurs, sometimes even when the sense has become otherwise morbidly impaired.

The spectrum from retained images of sight are seen of a different tint or colour, according as the eyes, when turned away from looking at the object, are darkened or directed to a light surface. In the former case, the lights and shadows are the same as appeared at the time of regarding the object (*positive spectrum*) ; in the latter, they are the reverse (*negative spectrum*). If the object from which the impression has been derived is coloured, the negative spectrum is coloured also, but differently, thus :—If the eye be fixed on a red coloured object for some time, and then turned away from it, a spectrum of the object will continue to be seen ; but instead of a red, of a green colour. If, on the contrary, the object looked at be green, the spectrum will be red; again, if blue, the spectrum will be orange; if orange, the spectrum will be blue; if yellow, the spectrum will be violet; if violet, the spectrum will be yellow. From this it is seen that the colour of the negative spectrum is always that which, being added to the colour of the object looked at, makes up the sum of the prismatic colours, yellow, red, and blue, which, by their combination, form white light; hence the name *complementary*, which has been given to the colour of the negative spectrum.

SCOTOMATA, OR MUSCÆ VOLITANTES.

The common appearance of motes floating in the field of view is owing to particles, naturally existing in the eye, throwing their diffracted shadows on the retina, close in front of which they have their seat. The appearance of motes may, under certain conditions, be seen by any person; but usually no notice is taken of them, as they are indistinct, present themselves occasionally only, and are not troublesome. They are most distinctly seen, and are most troublesome, when there exists impaired energy of the retina. Such a state of irritable weakness of the retina may, therefore, be viewed as the general condition on which floating muscæ, considered as a disease, depend.

Another form of scotomata consists in the appearance of particles darting in every direction before the eyes, like the blood corpuscles circulating in the web of the frog's foot, as seen under the microscope. This, the circulatory spectrum, may also be seen, under certain conditions, by a healthy eye, though not so distinctly as in the morbid state.

FIXED SPECTRUM.

The appearance, like a vertical bar in the field of view, mentioned in one of Mr. Erichsen's *cases, was an example of a fixed spectrum. Fixed spectra are owing to insensible spots of the retina, and constitute a state of partial amaurosis.

IMPAIRMENT OF SIGHT.

The impairment of sight which accompanied, or was ushered in by, the symptoms just described, occurred in some of the cases as a transitory haze only at first, but usually the dimness was, or became, persistent. It affected one eye only, or both. In the latter case one eye was found dimmer than the other.

The persistent dimness, or uniform feebleness of sight, is owing to a commencing atrophy of the optic nerve.

Transitory failure of sight is owing, as shall be more particularly explained below, to a transitory aggravation of the already disturbed state of the circulation in the optic nervous apparatus, viz., stoppage of the afflux of arterial blood in consequence of temporary constriction of the arteries.

As it is indifferent what part of the optic nervous apparatus be excited, in order that luminous spectra may be perceived, so, in such case of failure of sight, the deterioration of nervous structure may have its seat first in the cerebral part of the optic nervous apparatus, or in the optic nerve, before the structure of the disc and retina is appreciably affected.

DIPLOPIA, OR DOUBLE VISION WITH THE TWO EYES OPEN.

In double vision with two eyes, the two images of the object looked at are seen at some distance apart, and not overlapping each other. On closing one eye the object is seen single.

Double vision with two eyes is the result of mis-direction of one or other eye, owing, most frequently, to paralysis of some one or more of the muscles moving it. Such misdirection is objectively evident in the form of a squint, if it be owing to paralysis of any of the recti muscles; but is not so when owing to abnormal action of the oblique muscles. In this case the misdirection consists in loss of parallelism of the vertical and horizontal diameters of the eyeballs. The relative position of the two images depends upon the direction and degree of the deviation of the eyes. Thus, when one eye is turned inwards, its image is on the same side; when turned outwards, it is on the opposite side. Again, the two images are seen side by side at a greater or less distance from each other when the deviation depends on paralysis of the internal or external rectus muscle; one above the other when the deviation depends on a non-correspondence in the action of the superior or inferior rectus muscle; one image appearing obliquely placed, and higher in respect to the other when the superior oblique muscle is paralysed.

In *polyopia,* or manifold vision with a single eye, no defect of correspondence in the direction of the two eyeballs is implied. To a single eye, or to both at the same time indifferently, objects appear not double merely, but manifold. The images, however, are not seen apart, but the appearance is as if several shadows were overlapping each other.

FAILURE OF THE POWER TO ESTIMATE DISTANCE AND POSITION, AND TO CO-ORDINATE VOLUNTARY MOVEMENTS.

By the combined action of the two eyes we are enabled to perceive at one glance the length, breadth, and thickness of near objects. The stereoscope shows this. The faculty of perceiving the three dimensions of space with the two eyes implies the faculty of recognising the distance and position of near objects by the same means. If one eye be misdirected, or its sight lost, we can no longer perceive position and distance so as to be able accurately to pour wine into a glass, snuff a candle, or perform the like operation.

Such mistakes, arising from stereoscopic defect of sight, are not to be confounded with loss of the power of guiding the voluntary movements, arising from impairment or loss of tactile sensibility, such as was exhibited in the following case related by Sir Charles Bell :—

CASE X.

A mother, while nursing her infant, was seized with a paralysis, attended by loss of muscular power on one side, and loss of sensibility on the other. Though she could hold her child to her bosom with the arm which retained muscular power, it was only so long as she kept her eyes on the infant.

Such a case, again, is different from what occurs in locomotor ataxy. Here, although the sight, as observed by Dr. Bazire, in his "Translation of Trousseau's Clinical Lectures," will, to a certain extent, make up for the instinctive control which the patient has lost over his movements, it is not sufficient to enable him to maintain the effective co-ordination which was manifested in the above case recorded by Sir Charles Bell, and, of course, still less the harmonious co-ordination of health. The loss of the power of guiding the voluntary movements while the eyes are shut has been observed by Dr. Brown-Sequard,* as respects the lower limbs, in cases of tumour pressing upon the extremity of the spinal cord on its posterior surface. The atrophy, thereby caused, of some of the posterior roots of nerves, and of the posterior white and grey parts of the spinal cord, produced partial anæsthesia of the skin and muscles of the feet and legs.

* "Lectures on the Diagnosis and Treatment of the principal forms of Paralysis of the Lower Extremities, 1861," p. 94.

PAINS AND NOISES IN THE HEAD, OVER-SENSI-
TIVENESS OF THE EARS TO SOUNDS, AND
DULLNESS OF HEARING.

Noises in the head are analogous in their nature and
cause to luminous spectra, while over-sensitiveness of
the ears to sound is analogous to intolerance of light,
and the dullness of hearing to impaired sight.

VERTIGO, OR DIZZINESS.

Whirling oneself round, it is well known, excites
the sensations of vertigo. Vertigo is also experienced
by most persons on looking down from a height.
Edgar, in " King Lear," says,—

> " —— How fearful
> And dizzy 'tis, to cast one's eyes so low !
> —— I'll look no more,
> Lest my brain turn, and the deficient sight
> Topple down headlong."

This is owing partly to the alarm and partly to the
oscillating of the eyeballs in the impossible effort to
combine stereoscopically the very dissimilar impressions
made by the abyss on the two retinæ.

Some persons are more liable to vertigo than others,
and in disturbance of the brain this tendency is morbidly increased, as was the case with the patient who
imagined that the carriage in which he was driving
was always in danger of running into the ditch or
hedge on the near side.

Vertigo usually begins in the sense of sight with the
sensation as if objects were moving in a circle or in

curves, or as if they alternately moved up and down, or oscillated in a transverse direction. When ocular spectra are present—luminous spectra, for example, as in a case to be related below—the subjective appearances are seen as if moving in the same manner. To the swimming of the sight succeeds the sensation as if the patient's own body followed the visual movements. Hearing next becomes affected with confusion and noises in the head. A trickling sensation of heat and cold over the body, numbness of feeling, and the sensation as if the ground were heaving under the feet ensue, with sickness at stomach and retching.

Anæmic persons when they stand up are liable to be seized with vertigo, as are plethoric persons when they lie with their heads low.

In vertigo there is, as regards sight, a failure of the command over the muscles of the eye necessary to fix the objects looked at ; as regards hearing, a similar failure over the adjusting muscles of the ear, and as regards touch, a failure of co-ordinating power over the movements of the limbs.

CHAPTER IV.

A HALF-CLOSED state of the eyelids—a sunken, dull, and watery aspect of the eyes—enlargement of veins on the white of the eyeballs, indicating choroidal congestion—such were some of the appearances observed. They will be noticed below. But the points claiming chief attention here are the state of the pupil, and the appearance of the fundus of the eye, as seen under the ophthalmoscope.

STATE OF THE PUPIL.

In some of the cases the movements of the pupil under the influence of light and shade appeared to be undisturbed; in some, the movements were rather more active than natural, in others rather more sluggish; in some, again, the pupil, after contracting on exposure to the light, or on converging the eyes on a near object, showed a tendency to fall back into the middle state of dilatation. In none of the cases was the pupil much dilated, but in some it remained wide enough for the ophthalmoscopic examination.   In other cases the pupil contracted so much under the stimulus of the light reflected into the eye that it became necessary to bring it under the influence of atropia, before the ophthalmoscope could be used.   Atropia produced its effect, for the most part, in the ordinary length of time.   The state of the pupil will come to be further considered when inquiring into the nature of the connection between the spinal injury on the one hand, and the affection of the eyes and sight on the other.

STATE OF THE FUNDUS OF THE EYE.

Increased vascularity and some degree of whiteness of the optic disc ; congestion of the retina ; blackish discoloration of the retina adjacent to the disc from pigmentous deposit ; pigmentous deposit in the disc itself.   Such were appearances observed in various degrees under the ophthalmoscope in   the   cases related.   These appearances, however, are not to be viewed   as   the   expression   of   the   whole   damage suffered by the optic nervous apparatus, nor as the sole expression of the morbid anatomical state on which the failure of sight and other subjective symptoms depended.

In some cases, indeed, there may not at first, as already stated, be any morbid condition of the optic disc and retina which can be distinctly perceived under the ophthalmoscope.   The morbid state on which the defect of sight and other subjective symptoms depend, may be at first confined to some central part of the optic nervous apparatus,* and ophthalmoscopic evidences of implication of the optic disc and retina may not present themselves until the case is in a more advanced stage. And then, the morbid degeneration of the optic disc and retina which may be observable, though a very efficient cause of impaired sight, is not to be viewed as the anatomical character of the amaurotic symptoms, because the amaurotic symptoms existed before the morbid degeneration in question presented itself, but rather as the effect, or, at least, as the coincident effect, of the morbid action which was going on in the intracranial part of the optic nervous apparatus, and which was the original cause of the failure of sight and its attendant symptoms.

If the morbid appearances at the fundus of the eyeball had been more pronounced than they were found to be at first, in the cases related, the failure of sight and the other subjective symptoms ought to have been in a correspondingly greater degree than appeared to be the case from the accounts given by the patients. With whiteness of the optic disc or pigmentous discoloration of the retina, amaurotic symptoms are unmistakeably pronounced. But there may be already very serious defect of sight before such alterations of structure, as are implied by the appearances named,

---

* Under the expression "optic nervous apparatus" are comprised : the corpora quadrigemina, the tractus optici, the commissure, the optic nerves, the optic discs and retinæ. Of these parts, the corpora quadrigemina, tractus optici, and commissure, are intracranial or central ; the optic nerves, intraorbital ; the discs and retinæ, intraocular or peripheral.

present themselves under the ophthalmoscope — a defect of sight which is owing to a slowly progressing inflammatory process, propagating itself from the intra-cranial part of the optic nervous apparatus peripherally along the optic nerve to the disc.

Injury or disease of a nerve at its root or in its trunk, let it be observed, is followed by atrophy of the peripheral termination of the nerve in the organ to which it belongs or is distributed. But this atrophy is not appreciable objectively for some time. The injury or disease of the nerve being at first manifested solely by the failure of power and other subjective symptoms. What is here said is well illustrated by the experiment of cutting the ischiatic nerve in a frog. The first and only appreciable effect of this injury is paralysis of the leg as regards both motion and sensation. Under the microscope no change can be discovered in the peri-pheral ramifications of the nerve in the web. Some time afterwards, however, on re-examination of the web, granular degeneration of the medullary contents of the nerve fibrils accompanying the arteries, indica-tive of commencing atrophy, will be discovered. White atrophy of the optic disc from implication of the root of the optic nerves in cerebral disease, is an analogous example of the result of this kind of degeneration of nerve fibres.

The white atrophy of the optic disc in the cases of failure of sight from spinal affection which we have under consideration, is of a different character. It arises from optic neuritis, and does not attain so complete a development as the white atrophy of cerebral amaurosis—at least for a very long time.

In this case the failure of sight proceeds *pari passu* with the alteration of structure, whereas in the case of destruction of the root of the optic nerve by injury or disease, white atrophy follows the loss of sight.

The state of the fundus will be considered in further detail, when the connection between the injury of the spinal cord on the one hand, and the affection of the eyes and sight on the other, comes to be inquired into.

CHAPTER V.

INQUIRY INTO THE NATURE OF THE MORBID CONDITION
OF THE OPTIC NERVOUS APPARATUS, WHICH WAS
THE PROXIMATE CAUSE OF THE AMAUROTIC SYMP-
TOMS IN THE CASES UNDER CONSIDERATION.

NATURE AND SEAT OF THE MORBID CONDITIONS
ON WHICH AMAUROTIC AFFECTIONS DEPEND.

THE perverted, impaired, or lost sensibility of the
optic nervous apparatus in amaurotic affections, we
know, may be the result of morbid conditions differing
both as regards nature and seat. As regards nature,
the morbid conditions may be congestion, hyperæmic
or anæmic, or actual inflammation and its consequences,
leading to degeneration of structure and nervous
exhaustion. Pressure by neighbouring parts in a
swollen state or by tumours, may also be a cause.

As regards seat, this may be, as has been said, in the
retina, or the optic nerve, or the cerebral portion of the
optic nervous apparatus. The retina, the optic nerve,
or that part of the brain with which the optic nerve is
connected, may be, together or separately, the seat of
the morbid condition on which the defect of sight
depends. If the retina only be affected, it cannot receive the impression which should be transmitted by

the optic nerve to the brain; if the optic nerve only be affected, it cannot transmit to the brain the visual impression received by the retina; if that part of the brain with which the optic nerve is connected be alone affected, the sensorial power to take cognisance of the visual impressions transmitted by the optic nerve from the retina is lost. The general result is, therefore, the same, whether the different parts of the optic nervous apparatus be affected all together, or separately at first.

Eventually affection of one part leads to degeneration of the whole; thus affection of the cerebral part of the optic nervous apparatus leads, by propagation of morbid action from centre to periphery, to white atrophy of the optic disc; whilst, *vice versâ*, a degenerated state of the eyeball leads, by propagation of morbid action from periphery to centre, to atrophy of the intracranial part of the optic nervous apparatus.

PARTICULAR NATURE AND SEAT OF THE MORBID STATE ON WHICH THE AMAUROTIC FAILURE OF SIGHT IN THE CASES UNDER CONSIDERATION DEPENDS.

From a review of all the phenomena, it would appear that in the class of cases we have under consideration, viz., failure of sight from spinal injury, the morbid state acting as the immediate cause of the amaurotic symptoms is referable to a disturbance of the circulation in the optic nervous apparatus.

The failure of sight, it is to be remarked, was not experienced until some considerable time after the

injury. This would show that the affection of the eyes
arose not from any immediate lesion, but from the loss
of vital energy and from impaired nutrition of the optic
nervous apparatus, occasioned by the disturbance of the
circulation,—impaired nutrition leading to degeneration
of structure. Altogether it resembles very much that
form of idiopathic failure of sight known under the
name of Erethitic Amblyopia or Amaurosis.

In a case of erethitic amblyopia or amaurosis of
idiopathic origin, the patient is unable to exert what
sight he has except for a few minutes, the eyes becoming
red from vascular injection, watery, unsteady, and painful, whilst the objects looked at appear still more indistinct and confused. Occasionally a haze transitorily
obscures the sight altogether. Scotomata, either in the
common form of muscæ volitantes or in the form of
the circulatory spectrum, glimmer in the field of view.
Luminous spectra often flash before the eyes in the
dark. The images of objects continue to be seen for an
undue length of time after the patient has turned away
from looking at the objects themselves. Sight is worse
in the morning, better in the course of the day, and
improved under a cheerful state of mind.

The pupils are often unduly active, contracting
greatly on exposure of the eyes to light and dilating
much in weak light, perhaps tending to dilate after the
contraction on first exposure to the light. The conjunctiva is pale, and pervaded with enlarged tortuous
vessels. Eyes weak and watery looking.

Under the ophthalmoscope the disc is seen to be

whitish and somewhat congested. The retinal veins
are large, though the fundus generally presents an
anæmic aspect, with perhaps some pigmentous degeneration of the retina round the disc.

Along with the affection of the sight there may be
vertigo, tinnitus aurium, fixed pain in the head, confusion of mind and muscular weakness, paleness of the
face and visible parts of the mucous membranes.

Erethism, or irritable weakness, is a manifestation of
impaired energy of the central nervous system from an
anæmic state of the brain, induced by loss of blood or
other debilitating influences, or some more local cause.

This state of the nervous system gives rise to unregulated constrictions and dilatations of the small
arteries of the optic nervous apparatus, whereby disturbance of the circulation is caused with impaired
nutrition. And this seems to be the condition, speaking
generally, on which the failure of sight depends.

I have already observed that the affection of the
sight, though not less serious in degree, has in some
cases been comparatively secondary in importance,
owing to the gravity of the injuries sustained by other
parts of the body. The symptoms of such other
injuries are analogous, many of them, in their nature,
to those we have studied in the optic nervous apparatus,
and equally with them point to a profound impression
on the great nervous centre having been made by the
shock. Into the particular nature of this we now
proceed to inquire.

CHAPTER VI.

MANNER IN WHICH THE SPINAL CORD IS AFFECTED IN CASES OF CONCUSSION OF THE SPINE.

PREPARATORY to inquiring into the special nature of the connection between the spinal concussion on the one hand and the affection of the eyes and sight on the other, it will be necessary to ascertain the manner in which the spinal cord itself is affected.

As a result of injury of the spinal cord, inflammation of its proper substance (*Spinal Myelitis*), or that of its membranes (*Spinal Meningitis*), is prone to take place.

SPINAL MYELITIS.

In spinal myelitis, the morbid appearance most commonly met with in *post-mortem* examinations is softening of the substance of the cord. This inflammatory softening may, according to Ollivier,* occupy very varying extents of its tissue. Sometimes the whole thickness of the cord is affected at one point; sometimes one of the lateral halves in a vertical direction is affected; at other times it is most marked

* Ollivier d'Angers—Traité des Maladies de la Moelle Epinière. Paris. 1837.

in, or wholly confined to its anterior or its posterior
aspect, or the grey central portion may be more affected
than the circumferential part. Then, again, these
changes of structure may be limited to one region
only,—to the cervical, the dorsal, or the lumbar. It is
very rare indeed that the whole length of the cord is
affected. The most common seat of the inflammatory
softening is the lumbar region; next in order of frequency, the cervical. In very chronic cases of myelitis,
the whole of the nervous substance disappears, and
nothing but connective tissue remains at the part
affected.

In spinal myelitis the grey substance of the cord is
almost always implicated. In the rare cases in which
the inflammatory softening is confined to the white
substance, the myelitis has been consecutive to inflammation of the meninges of the cord.

The symptoms of spinal myelitis are referable to
three heads, viz., morbid sensations, irregular movements, and alteration of temperature. And according
to the particular part of the cord which is the seat of
the inflammation, so is the place where the morbid
sensations are felt, where the irregular movements are
manifested, and where the temperature is altered.

According to differences in the seat of the inflammation in the spinal cord, observes Dr. Brown-
Sequard,* there are great differences in the intensity
of the symptoms, and still more, some of them may be
missing. When the grey matter of the spinal cord is

* Opere citato, ut supra, p. 33.

the seat of the inflammation, all the characteristic symptoms exist, and to a notable degree; but in the rare cases in which the inflammation is limited to the white columns of the spinal cord, the symptoms present a marked difference from those which occur when the grey substance is implicated. If the posterior columns be the seat of inflammation, the symptoms are much less prominent, but they all exist. When the disease is in the anterior columns there is no *anæsthesia* or loss of sensibility, and morbid sensations referred to the paralysed parts of the body hardly exist.

Morbid Sensations in Spinal Myelitis.—Pain of the spine, either spontaneous or on pressure over the spinous processes at the place corresponding to the part of the cord which is the seat of the inflammation. Sensations of tightness or pressure round the body. Formication, or creeping. Pricking and tingling in the fingers and toes. Sensation of burning or coldness. The parts otherwise little or not at all sensible to external impressions. When the grey substance of the cord is not implicated, this anæsthesia does not exist.

In non-inflammatory softening there is no sensation of tightness round the body nor other referred sensations.

Irregular Movements in Spinal Myelitis.—The peculiar kind of sensibility on which the instinctive co-ordination of the muscular movements depends is impaired from the very first. Spasmodic movements, cramps, and twitchings in the body and lower limbs occur in fits.

Spinal myelitis is the most common cause of paralysis of the lower limbs. White, or non-inflammatory softening, says Dr. Brown-Sequard, is the next most common cause ; but along with the paraplegia in this case, there are no spasms, cramps, nor twitchings.

Alteration of Objective Temperature in Spinal Myelitis.—In paraplegia from myelitis, the objective temperature is generally below the natural standard in consequence of the irritation to which the roots of the vaso-motor nerves are exposed. The effect of this irritation is constriction of the small arteries, whereby the afflux of arterial blood to the parts affected is diminished, and along with this the supply of heat.

In non-inflammatory softening, the temperature of the paralysed limbs is usually more elevated than in health, from paralysis of the vaso-motor nerves and consequent dilatation of the arteries, with increased afflux of blood.

SPINAL MENINGITIS.

In spinal meningitis, the meningo-rachidian veins are turgid with blood, and the vessels of the pia-mater much injected—sometimes in patches, at others uniformly so. Red serous fluid, clear or opaque, from the admixture of lymph, has been found effused in the cavity of the arachnoid. Ollivier* states that one of the most constant appearances in chronic spinal meningitis is adhesion between the serous lamina lining the dura-mater, and that which invests the spinal pia-mater. This, he says, he has often observed, and

* Op. cit., ut supra.

especially in that form of the inflammation which is developed as the result of a lesion of the vertebræ. Lymph of puriform appearance has also been found under the arachnoid, between it and the pia-mater.

Ollivier also says that it is rare to find meningeal inflammation limited to the membranes in the vertebral canal. That we see at the same time a more or less intense cerebral meningitis. In the cases that he relates of spinal meningitis, he makes frequent reference to co-existing cerebral symptoms—states that they often complicate the case so as to render the diagnosis difficult, especially in the early stages. In the *post-mortem* appearances that he details of patients who have died of spinal meningitis, he describes the morbid conditions met with in the cranium as indicative of increased vascularity and inflammation of the arachnoid. In both the fatal cases of meningitis of the spine recorded by Abercrombie evidences of intracranial mischief are described.

The cerebral symptoms pointing to chronic basilar meningitis are :—Pains in the head ; pain and tension in the nape of the neck ; disturbance of the mental and sensorial functions ; restless sleep ; convulsions ; pale and depressed aspect ; and eventually a break up of the whole system.

It is worthy of remark that in certain of the cases above related in which the sight and hearing were more especially affected on one side, it was in the limbs of the opposite side that the paralytic symptoms were most developed. From this it may be inferred that the

E

accompanying impairment of locomotive power was in part owing to the supervening cerebral implication.

Morbid Sensations in Spinal Meningitis.—Pain and stiffness like rheumatism in the back, greatly aggravated by movement of any kind, whether of the spine or lower limbs; but no pain on pressure over the spinous processes as in myelitis. Acute pains, extending from the spine to the lower extremities, occur of themselves also.

There is sometimes hyperæsthesia, or over-sensibility, but rarely anæsthesia or loss of sensibility.

Irregular Movements in Spinal Meningitis.—Rigid spasm of the muscles of the back induced by the pain on moving the spine. When this spasm is in the cervical region the head is bent back; when it is all along the spine a state like oposthotonos presents itself. Spasmodic movements and cramps in the lower limbs also sometimes occur.

Paraplegia is variable.

Alteration of Temperature in Spinal Meningitis.— In meningitis, the temperature of the affected limbs may be below the natural standard, but like the paraplegia, it is variable; and is more circumscribed as regards the parts implicated.

SPINAL MYELITIS AND MENINGITIS COMBINED.

It is important to observe, Mr. Erichsen remarks, that, although spinal meningitis and myelitis are occasionally met with, distinct and separate from each other, they most frequently co-exist. When thus co-existing, and even when arising from the same cause, they may be associated with each other in very varying degrees ; in some cases the symptoms of meningitis, in others those of myelitis, being the most marked. After death the characteristic morbid-anatomical appearances present a prominence corresponding to that exhibited by the symptoms during life.*

CONGESTION OF THE SPINAL CORD AND MENINGES.

In spinal congestion the symptoms above enumerated of myelitis or meningitis occur as regards sensation, motion, and temperature, though in a less degree.

The symptoms are usually worse on rising in the morning.

SIGNIFICATION OF THE SYMPTOMS IN SPINAL MYELITIS, MENINGITIS AND CONGESTION.

Morbid sensations.—The referred sensations of heat or cold, formication, tickling, pressure, tightness, pins and needles, failure of muscular sense, are owing to the irritation of the grey substance of the cord, and also, though in a less degree, to the irritation of the posterior roots of the nerves implicated with the affected spinal membranes. In myelitis, Dr. Brown-Sequard remarks,† the sensations may be referred to all parts receiving their sensitive nerves from the cord below the upper limit of the inflammation ; whereas in meningitis and congestion, the sensations

---

* Op. cit. p. 118.　　　　† Op. cit.

are referred to parts receiving their nerves from the cord at the seat only of the irritation.

There are no referred sensations in cases of non-inflammatory softening of the cord.

Irregular movements.— Cramps, twitchings, and other irregular movements are owing to irritation of the anterior or motor roots of the spinal nerves, or to irritation of the spinal cord at their roots, or to excitement of the motor nerves by reflex action from the irritated posterior roots. Such symptoms are common in myelitis. Rigid spasm in the back indicates spinal meningitis.

Twitchings are more frequent than cramps in chronic meningitis or congestion.

In myelitis the sphere over which paralysis extends, and the frequency of cramps, is greater than in congestion and meningitis, for in myelitis the number of motor conductors implicated is larger and the irritation greater.

The chief cause of paralysis in congestion and meningitis being pressure on the nerves in their passage from the spinal canal, and this pressure being variable and limited to a few nerves, the degree of paralysis is variable and its sphere circumscribed.

There are no irregular movements in white softening.

Alteration of temperature.—Objective diminution of temperature indicates irritation and excitement of the roots of the vaso-motor nerves in the part of the spinal cord affected. Objective elevation of temperature, on the contrary, indicates paralysis of the vaso-motor nerves by injury of their roots, as in non-inflammatory softening of the cord.

SIGNIFICATION OF THE SYMPTOMS OF CONCUSSION OF THE SPINE.

Having thus detailed the symptoms commonly recog nised as connected with and dependent upon spinal myelitis, meningitis, and congestion in general, we have now to consider in a particular manner the signification of the symptoms that are described as characteristic of concussion of the spine from slight injuries and general shocks of the body. This I shall do with the assistance of Mr. Erichsen, who has discussed the subject in so lucid a manner.

In those cases, says Mr. Erichsen,[*] in which the shock to the system has been general and unconnected with any local and direct implication of the spinal column by external violence, and where the symptoms are less those of paralysis than of disordered nervous action, the pathological states on which these symptoms are dependent are of a more chronic and less directly obvious character than are the pathological states in those cases that result from direct and violent blows upon the back. They doubtless consist mainly of chronic and subacute inflammatory action in the spinal membranes and in chronic myelitis, with those changes in the structure of the cord that are the inevitable consequences of a long-continued chronic inflammatory condition developed by it.

* Op. cit. ut supra.

Mr. Erichsen gives the result of a *post mortem* examination of the spinal cord of a person who had died from the remote effects of concussion of the spine, supplied to him by Mr. Gore, of Bath. The patient, a middle-aged man of active business habits, had been in a railway collision, and, without any signs of external injury, fracture, dislocation, wound, or bruise, began to manifest the usual nervous symptoms. He gradually, but very slowly, became partially paralysed in the lower extremities, and died three years and a half after the accident.

On the occurrence of the collision, the patient walked from the train to the station close at hand; though he had no external sign of having received any injury, he complained of a pain in his back. Being most unwilling to give in, he made every effort to get about on his business, and did so for a short time after the accident, though with much distress. Numbness and a want of power in the muscles of the lower limbs gradually but steadily increasing, he at last became disabled. There was much sensitiveness to external impressions, so that a shock against a table or chair caused great distress. As Mr. Gore did not see the patient until about a year after the accident, and, thereafter, only at intervals up to the time of death, he was not able to inform Mr. Erichsen of the precise time when the paralytic symptoms appeared. He says, however, that it was certainly within less than a year of the time when the accident occurred. In the latter part of the patient's illness, some failure of the sensi-

bility on which the co-ordination of the muscular movements depends became apparent in the upper extremities, so that if he was off his guard a cup or a glass
would slip from his fingers.* There was no paralysis
of the sphincter of the bladder until about eighteen
months before his death, when the urine became pale
and alkaline with muco-purulent deposits. In this case
the symptoms were not so severe as usual, there was
no very marked tenderness or rigidity of the spine, nor
were there any convulsive movements.

On examination after death, traces of chronic inflammation were found in the arachnoid and the cortical
substance of the brain. The spinal meninges were
greatly congested, and exudative matter had been
deposited upon the surface of the cord. The cord itself
was much narrowed in its antero-posterior diameter in
the cervico-dorsal region. The narrowing was owing to
absorption of the posterior columns. These had not
only to a great extent disappeared, but the remains
were of a dark-brownish colour, and had undergone
important structural changes.

As effects of " Concussion of the Spine," from a railway accident, this case presents evidences of chronic
meningitis—cerebral as well as spinal—of chronic
myelitis, with subsequent atrophy of the posterior
columns.

The spinal symptoms that occurred in the cases of
" Concussion of the Spine," which have been related,
continues Mr. Erichsen, consisted briefly in pain at one

* See above, pp. 32—33.

or more points of the spine, greatly increased on pressure, and on movement of any kind, so as to occasion extreme rigidity of the vertebral column.

Ollivier says that one of the most characteristic signs of spinal meningitis is pain in the spine, which is most intense opposite the seat of inflammation. This pain is greatly increased by movement of any kind, so that the patient fearing the slightest displacement of the spine, preserves it in a state of absolute quiescence. This pain is usually accompanied by muscular rigidity. It remits, sometimes being much more severe than at others, and occasionally even disappears entirely. The pain of spinal meningitis is not increased by pressure. In chronic myelitis, there is a painful spot in the spine, where the pain is increased on pressure, and this is looked upon as indicative of inflammation of the cord rather than of the membranes.

The symptoms dependent on concussion of the spine referable to the limbs, may briefly be stated to consist in painful sensations along the course of the nerves, followed by more or less numbness, tingling and creeping; some loss of co-ordinating motor power in one or more of the limbs, giving rise to peculiarity and unsteadiness of gait.

The sphincters are not paralysed.

These are the very symptoms that are given by Ollivier and others as characteristic of spinal meningitis, but more particularly of myelitis.

In spinal meningitis, says Ollivier, there is increased sensibility in different parts of the limbs, extending along the course of the nerves, and augmented by the most superficial pressure. These pains are often at first mistaken for rheumatism. There is often also more or less contraction of the muscles.

In myelitis the sensibility is at first augmented, but after a time becomes lessened ; and the anæsthesia is accompanied by various uneasy sensations in the limbs, such as formications and a feeling as if the limb was asleep. These sensations are first experienced in the fingers and toes, and thence extend upwards along the limbs.

Some degree of paralysis of movement occurs in certain sets of muscles—or in one limb. Thus the lower limbs may be singly or successively affected before the upper extremities, or *vice versâ*. Occasionally this loss of power assumes a hemiplegic form. All this will vary according to the seat and the extent of the myelitis.

In chronic myelitis patients often complain of a sensation as of a cord tied tightly round the body.

The gait of patients affected with chronic myelitis, Ollivier remarks, is peculiar. The foot is raised with difficulty, the toes are sometimes depressed and at others they are raised, and the heel drags in walking. The body is kept erect and carried somewhat backwards.

If we take any one symptom that enters into the composition of these various groups, Mr. Erichsen concludes, we shall find that it is more or less common to various forms of disease of the nervous system. But if

we compare the groups of symptoms that have just
been detailed, their progressive development and indefinite continuance, with those which are described by
Ollivier and other writers of acknowledged authority
on diseases of the nervous system, as characteristic of
spinal meningitis and myelitis, we shall find that they
mostly correspond with one another in every particular
—so closely, indeed, as to leave no doubt that the
whole train of nervous phenomena arising from shakes
and jars of, or blows on, the body, and described as
characteristic of so-called " Concussion of the Spine,"
are in reality due to chronic inflammation of the spinal
membranes and cord.   The variation in different cases
being referable partly to whether meningitis or myelitis
predominates, and in a great measure to the exact
situation and extent of the intraspinal inflammation,
and to the degree to which its resulting structural
changes may have developed themselves in the membranes or cord.

CHAPTER VII.

CHRONIC inflammation of the spinal cord and its meninges appears, from Mr. Erichsen's lucid pathological exposition of the subject, quoted in the preceding chapter, to be the morbid state which is first excited by spinal concussion. Whether the cerebral disturbance, including the affection of the optic nervous apparatus, &c., which has been found so often to supervene, arises by extension of the inflammation, through continuity of structure, from the spinal cord to the brain, or whether, still supposing the spine to be *fons et origo mali*, it arises only by transmission to the brain of morbid action through an indirect channel,—is the question which here presents itself in our inquiry into the nature of the connection between the spinal concussion on the one hand and the affection of the eyes and sight on the other.

The cerebral disturbance with which the patient is often found to be affected immediately after the accident is, no doubt, owing to concussion having at the same time acted on the head; but cases of this kind are here excluded from present consideration.

The supervention of a disturbance of the circulation in the intracranial part of the optic nervous apparatus causing the affection of the eyes and sight, can scarcely be attributed to a propagation of congestion or inflammation from the spinal cord through the medulla oblongata to the brain; for the morbid condition of this important part of the central organ of the nervous system, which such a propagation of morbid action implies, would probably be attended by more immediately serious symptoms than those which do present themselves.

In the following case, for the opportunity of seeing which I am indebted to Mr. Erichsen, it can scarcely, I think, be doubted that the connection between the affection of the spinal cord, and that of the optic nervous apparatus, was not by extension of inflammation from the spinal cord through the continuous structure of the intervening medulla oblongata to the brain, but by transmission of morbid action from the spinal cord to the brain through an indirect channel.

CASE XI.

Mrs. K., æt. 29, is believed to have suffered an injury of the cervical spine, when about eighteen months' old, from a fall out of her cot. Nothing certain, however, is known of the accident, as the infant was not immediately under her mother's eye at the time, but under the care of a nurse. It is, nevertheless, certain that the patient, without being otherwise ill, has been restless and uneasy from childhood. Though very talented and accomplished, she was never able to apply herself to her studies in a steady and continuous manner.

Certain of the eye symptoms were those to which attention was first drawn, and it was not until some time after that the patient became conscious of anything the matter with her spine. There can be no doubt, however, that the spinal irritation was the starting point of all the symptoms.

The eye symptoms referred to came on about ten or twelve years ago, and consisted of a sensation, as the patient describes it, of rubbing at the back of the eye, followed next day by the vertiginous appearance, as if objects were gliding past her. She had then no pain. The patient still refers to the back of the eyes as the seat of the disturbance causing the distressing visual symptoms to be described below.

Mr. Erichsen, in his narrative of the case, published three years ago (Op. cit. p. 61), states that he found a distinct projection backwards of the spinous processes of the fifth and sixth cervical vertebræ, where the patient complained of a constant pressure and pain of a grating or grinding character, as if the bones were in contact with each other. From this point a peculiar sense of uneasiness spread itself over the whole of the body and limbs, producing tingling sensations and pains of the most distressing character.

The patient can walk well so long as there is anything near her. Thus, she can walk along a street guided by the area railings, but when she comes to an open space, such as a square or crossing, she is lost, and requires to be guided, else she would fall.

The hands and feet are always cold, even in summer.

At the time Mr. Erichsen wrote the sight was strong, but she saw the circulatory spectrum before her eyes, the blood corpuscles apparently spinning round in convolutions, and often coloured. The hearing was good, but she had unceasing loud noises in her head, like "gravel-stones" rolling through it. No perversion of smell or taste existed.

When I saw Mrs. K. on the 16th of January, 1869, she complained that wherever she looks, objects appear as if in motion, heaving all about, and that luminous spectra are almost constantly seen whirling or revolving before her eyes. Sometimes, instead of the luminous spectra, a dark shade comes over the sight.

Mrs. K. has no rest from these sensations of vertigo day or night. Sometimes nausea and sickness at stomach are occasioned by the heaving appearances ; has little sleep, and what she has is disturbed by dreams of her distressing sensations.

The sight is much impaired now. The patient sitting near could not see my face distinctly enough to recognise my features. She could see only parts of my face, and they too appeared to her to have a heaving or undulating motion. Besides this, the effort to look at any thing excites an outbreak of the photopsy, or luminous spectra, which compels her to shut her eyes, so that one might at first imagine that she was labouring under photophobia, or intolerance of light.

Pressure on the neck, where the projection backwards of the spinous processes of the fifth and sixth cervical vertebræ exists, aggravates the pain there felt, and excites an extension of it to the eyes, and, at the same time, an outburst of the luminous spectra. On leaning the head backwards, even, a jar runs throughout the whole body, and photopsy is excited.

The distressing symptoms now described, have of late become much aggravated.

On examining the eyes, I found the pupils in a medium state of dilatation, and rather inactive. Under the ophthalmoscope, the optic discs were observed to be of a dull bluish-white in the temporal half, and not very well defined on the nasal side, in consequence of vascular congestion. The disc of the right eye was more altered than that of the left. The retinal vessels appeared small. The retina adjacent to the disc had a slight blackish discoloration, and was anæmic-looking. The eyes bore the ophthalmoscopic examination very well.

Besides being troubled with the noises in her ears, Mrs. K. was somewhat dull of hearing. On examining the ears, I found the tympanic membranes opaque, and the skin of the adjacent part of the walls of the auditory passages red, and somewhat abraded, like what we so often find when there has been a morbid accumulation of wax, as was the case here.

By the touch, objects are felt as if they were heaving in a manner analogous to that in which they appear to heave before the eyes.

Some three years ago, Mrs. K., then Miss B., married, in the hope that such a change of life might have the effect of relieving her complaint, but in vain. She has had one child—a fine child—which she nursed for three months, when her milk failed.

It is above remarked that there can be no doubt that the spine was, in this case, the starting point of all the symptoms. Such was the view already taken by Mr. Erichsen, who remarks (Op. cit. pp. 62-3) that at the place of excurvation in the cervical vertebræ there had evidently existed disease, leading to organic changes, to which the remarkable train of general phenomena presented by this case were, doubtless, referable. " If I were to hazard an opinion," continues Mr. Erichsen, " it would be that some thickening of the meninges of the cord had probably taken place, the effect of which was to interfere with the sensory portions of the cord, rather than with the motor." The additional symptoms which have manifested themselves since this was written, show that the roots of the sympathetic or vaso-motor nerves at the place have also become implicated.

THE SYMPATHETIC NERVE THE MEDIUM OF
TRANSMISSION OF MORBID ACTION FROM THE
SPINAL CORD TO THE OPTIC NERVOUS APPA-
RATUS.

Inquiring now into the nature of the connection between the affection of the spinal cord and that of the
optic nervous apparatus, the auditory nervous apparatus, and certain other parts of the brain, in the class
of cases we have under consideration, the conclusion I
have come to is : that the connection consists in transmission to the brain of morbid action through that part
of the sympathetic nerve which, arising from the cervico-
dorsal portion of the spinal cord, passes through the
rami-communicantes of the corresponding spinal nerves
to join the sympathetic in the neck, and thence enters
the cranium in the internal carotid and vertebral plexuses. The sympathetic nerve, it is well-known, supplies
the vaso-motor nerves which govern the contractions of
the muscular walls of the arteries, and thus regulate the
width of the bore of these vessels. The capillary circulation, and consequently, the nutritive process, in a part
is thereby influenced—promoted or retarded—according
to circumstances.

The roots of the portion of the sympathetic mentioned, being implicated in the affection of the spinal
cord, the influence which that nerve naturally exerts on
the contractions of the muscular walls of the internal
carotid and vertebral arteries, and their branches, is
morbidly interfered with, and the result is a disturbance

F

of the circulation, with its attendant abnormal nutrition, in the parts of the nervous system indicated, viz., the optic nervous apparatus, &c.

EFFECTS OF SPINAL INJURY ON THE SIGHT COMPARED WITH THOSE OF LOCOMOTOR ATAXY.

In the supervention of amaurotic failure of sight, as well as in other respects, *progressive locomotor ataxy* presents a remarkable similarity to the class of cases we have under consideration. An examination of the nature of that disease, therefore, will be useful here in illustrating the connection between the spinal injury and the supervening amaurotic failure of sight.

Trousseau* expresses the opinion that the *post-mortem* appearances found in progressive locomotor ataxy are not the cause but the *effect*, the *product*, of the disease. By this he means only those material lesions which can be detected by our present means of investigation. For, as he justly states, he cannot conceive a functional disturbance without a special corresponding modification of the organ which discharges that function. This may, indeed, be more or less transitory, and it frequently does not alter the structure of the organ.

When we examine the lesions found in locomotor ataxy, we are struck, continues Trousseau, with three facts. *Firstly*, the atrophy of the nerve-tissue of the posterior columns and the corresponding roots.

* Lectures on Clinical Medicine delivered at the Hôtel Dieu, Paris. Translated by P. Victor Bazire, M.D., Lond. and Paris. Published by the New Sydenham Society. 1868.

Secondly. The development of cellular tissue, or, if you prefer the term now generally adopted, hypertrophy of the neuroglia. *Thirdly.* The vascularity of the diseased tissues.

The atrophy of the nerve-substance is the most striking phenomenon. It is, however, only a conse-quence of the pathological *evolution* of the cellular element, which, by developing itself, has crushed the nerve-elements contained within its areolæ; and this abnormal development of the cellular tissue is itself dependent on the increased vascularity of the tissues.

Now, is this increased vascularity sufficient to cha-racterise inflammation, and are we, from its presence, to conclude that progressive locomotor ataxy is only *a variety of chronic myelitis?* If so, how are we to ex-plain why this myelitis is always so exactly limited to the posterior columns of the cord, and to the roots issuing from them, and particularly why, during life, it is attended with symptoms differing so much in their form, course and changeability, from those common to all varieties of myelitis ?

This abnormal vascularity of the posterior columns of the cord which is again observed in the *motor oculi* and *optic nerves,* and the *tubercula quadri-gemina,* seems to be a consequence of frequently repeated congestions, analogous to those which we see during the patient's life affecting the conjunctiva. This membrane, as already mentioned, becomes in-jected in the interval between the paroxysms of pain, simultaneously with the occurrence of contraction of the pupil which is sometimes carried to an extreme degree. Generally, however, this injection disappears on the supervention of pain, especially of pain in the head, whilst the pupil dilates more or less at the same time.

These congestive phenomena show themselves in other diseases acknowledged to belong to the class of neuroses, such as hysteria, asthma, and Graves' disease (ex-ophthalmic goître) ; and they belong, Trousseau is of opinion, to the same category as those which, in his experiments, Professor Cl. Bernard produces at will by dividing the sympathetic in the neck.

The important bearing which the sympathetic nerve in the neck and its vaso-motor branches thus have on the investigation of our subject, claims for them the special notice to which the following chapter is devoted.

CHAPTER VIII.

THE SYMPATHETIC NERVE IN THE NECK—ITS ROOTS
IN THE CERVICO-DORSAL REGION OF THE SPINAL
CORD, AND ITS BRANCHES TO THE INTERNAL
CAROTID AND VERTEBRAL PLEXUSES.

THE region of the spinal cord where the part of the
sympathetic nerve in question has its roots is, according to Dr. Brown-Sequard's experiments,* that which
extends from the sixth cervical down as far as the
ninth or tenth dorsal vertebra.

Wutzer † showed that the sympathetic derives
radicle fibres from both roots of the spinal nerves,
and Mayer ‡ traced these fibres in the roots of the
nerves as far as the spinal cord itself. From Professor
Budge's experiments,§ it would further appear that
the sympathetic fibrils which are connected with the
region of the spinal cord here indicated are, some of
them, *centrifugal* or motor, and some *centripetal* or

---

* Sur les Résultats de la Section et de la Galvanisation du Nerf
Grand Sympathetique au Cou. In *Gazette Médicale de Paris.* 1854
† Müller's Archiv, 1834, p. 306.
‡ Nova Acta, xvi., p. 2.
§ Ueber die Bewegung der Iris, Für Physiologen und Aerzte.
Braunschweig. 1855.

excitor; and that the former pass on their way from the spinal cord to join the sympathetic in the neck, through the anterior roots of the corresponding spinal nerves, whilst the latter pass through the posterior roots on their way to end in the spinal cord.

The ganglion-cells in the grey central substance of the spinal cord which have been regarded as the origin of the roots of the sympathetic nerve-fibrils, are those which lie near the centre between the cells of the anterior roots, and the cells of the posterior roots of the spinal nerves.

Anatomically, the fibrils under consideration, we have seen, may be traced along the *rami-communicantes* of the spinal nerves, in the form of grey fasciculi, from the sympathetic towards the spinal marrow. Within the intervertebral foramina, their connection with the anterior roots of each spinal nerve is very obvious; but with the posterior roots, not so much so, though filaments may be traced by dissection, back towards the ganglions there. Of course it is not by anatomy, but only by experiments, such as those which have been referred to, that the physiological endowments of the sympathetic fibrils connected with the anterior and posterior roots of the spinal nerves can be determined.

In following out the demonstration of the pathology of the cases we have under consideration, I beg now to recall briefly to mind the anatomy of the plexuses of the sympathetic and their offshoots which

accompany the carotid (external and internal) and
vertebral arteries.

Of the anterior branches of the superior cervical
ganglion, some form the external carotid plexus. The
internal carotid plexus is formed by the ascending
branches of the superior cervical ganglion, and gives
off the filaments which are distributed to the muscular
walls of its branches; amongst others, the arteries of
the optic nervous apparatus within the cranium and the
ophthalmic artery. From the carotid plexus also, are
derived the filaments which, having joined the ophthalmic division of the fifth nerve in the cavernous sinus,
enter the orbit. The nasal branch of the ophthalmic,
containing many of these sympathetic fibrils, gives off
the long root of the lenticular ganglion and a few
ciliary nerves; the majority of the ciliary nerves being
derived through the lenticular ganglion. What has
been called the middle root of the lenticular ganglion
is a filament from the carotid plexus, which joins that
body between its long and short roots, whilst the short
root of the ganglion is derived from the lower branch
of the oculo-motor nerve.

The ciliary nerves accompany the arteries of the
same name into the eyeball. The central artery of the
retina also, as was discovered by Tiedemann, is accompanied into the eye by a minute branch from the ciliary
ganglion.

The fibrils of the oculo-motor nerve, which the
ciliary nerves derive from the short root of the lenticular ganglion, govern the contractions of the circular

muscular fibres of the iris or *sphincter pupillæ* ; while the sympathetic fibrils, the origin of which has been above described, govern the contractions of the muscular walls (composed of circularly disposed fibres) of the arteries and of the radiating muscular fibres of the iris or *dilator pupillæ.*

Before proceeding further with our special investigation as to the influence of the sympathetic nerve in the neck on the circulation in the head and eyes, we must apply ourselves to the study of the manifestations of the implication of the roots of the vaso-motor nerves generally in spinal injury or disease, and the effects of that implication on the temperature, nutrition, and vital energy of parts. But preparatory to this, again, it will be necessary to premise chapters on the Phenomena of the Circulation in the extreme vessels and the influence of the vaso-motor nerves thereon.

CHAPTER IX.

GENERAL DESCRIPTION OF ARTERIES, CAPIL-LARIES, AND VEINS.

As an introduction to this chapter, it is necessary to premise some remarks on the capillary vessels, the arteries, and the veins.

Capillary vessels are the channels which intervene between the last ramifications of the arteries and the radicles of the veins. Their average width is such as to give easy passage to the blood-corpuscles in single file. In the web of the bat's wing or frog's foot under the microscope they are *never* observed to become constricted or dilated. It is inferred, therefore, that the walls of the capillaries are not endowed with the vital property of contractility. Anatomy shows that they are not muscular. The capillaries are, in fact, passive tubes; exerting no force either in the promotion or in the retardation of the passage of the blood through them. The absence of contractility of the walls of the

capillaries and their passiveness in the circulation I
have elsewhere insisted on, and would again, on the
present occasion, take the liberty to insist on, as every
now and then we read or hear of the "contraction of
the capillaries" and the "action of the capillaries," as
exerting a great influence in promoting the circulation
of the blood and on the inflammatory process.

I shall inquire below into the manner in which the
flow of blood appears to be promoted in those organs in
which the force of the heart may be insufficient.

Whilst the width of the capillaries does not undergo
any variation due to contractility of their walls, the
arteries are observed occasionally to vary in width—
becoming constricted (perhaps even to closure of their
calibre for the time), and after an indefinite interval
regaining their previous or even a greater width. These
variations in the width of arteries, it is to be particularly
remarked, are not rhythmical like the systole and
diastole of the heart, and are not to be confounded with
the arterial pulse and arterial tension.

The pulsation of the arteries of the web of the frog
may be observed under the microscope. I have counted
thirty-two pulsations in the minute. At each pulsation
the artery is seen to become slightly dilated, the tonicity
and elasticity of its wall yielding to the blood forced
into it at each stroke of the heart. If the artery be at
all tortuous, or undulating, it becomes, at the same
time, still more bent at the bendings. The increased
bending or *locomotion* of the artery is more readily
detected than the slight dilatation. The effect of the

systole of the heart on the small arteries in the conjunctiva of the human eye, as communicated by the arterial tension of the large trunks, is manifested by increased bending. I have noticed in the frog's web an artery, even when, in consequence of temporary constriction, no blood was flowing in it, bend more at the bendings at each stroke of the heart; hence it may be observed that such bendings, when presented by the retinal arteries viewed under the ophthalmoscope, are no evidence that the circulation is going on in those vessels.

The pulsations of an artery are observed to be less and less evident down to the capillaries. When an artery has been cut across, there is no longer any pulsation below the wound, but pulsation continues above.

The vital property of tonic contractility, with which the walls of the arteries are endowed, has its seat in their middle coat, which is composed of circularly disposed organic muscular fibres. The constriction of the arteries is produced by contraction, and their dilatation permitted by relaxation of this muscular coat.

That variations, though slight, in the width of the veins of the frog's web take place, is indicated by appearances such as the following, seen under the microscope :—First, veins may sometimes be observed with their diameter less, and their wall thicker, than similar veins usually are ; secondly, in anæmic frogs, in which, from the emptiness of the vessels, dilatation cannot be attributed to distension, veins may sometimes be seen unusually wide, with their wall thin, and presenting none of those slight constrictions, with thickening here and there, which under other circumstances are presented by the veins, and which are similar to, though in a very much less degree than, what may be presented by the accompanying arteries. The constrictions with thickening of the walls referred to, are owing to the action, as we must suppose, of muscular fibres contained in the coat of the veins, though few in number. By microscopical examination of the structure of the walls of veins taken from other parts of the body of the frog, it is found that there is a greater development of the muscular coat, but from the position of the vessels, we cannot observe them during life so as to see their constrictions take place. This, however, we can do in the rabbit's ear.

Viewing a live rabbit's ear spread out against the light, we can see that the veins undergo tonic constrictions; and by anatomico-microscopical examination of a piece of vein cut out from the ear of a dead rabbit, we find a pretty well developed muscular coat composed of circular fibres resembling those of the muscular coat of arteries. This muscular coat of the veins is the agent of the tonic contractions, by which the constrictions we observe these vessels to undergo are produced.

In 1851, I discovered that the veins of the bat's wing, which are furnished with valves, are endowed with rhythmical contractility like the heart, and that the onward flow of blood is accelerated by each systole or contraction. The rhythmical contractions of the veins of the bat's wing are considered at the end of this chapter.

THE EFFECTS OF CONSTRICTION AND DILATATION OF THE SMALL ARTERIES OF A PART ON THE CIRCULATION THEREIN.

Arteries being capable, by virtue of the tonic contractility of their walls, of varying in width, the fulness of the stream, and the rapidity of the flow of the blood in them is subject to occasional variation, independently of any variation in the force of the heart's action. In the web of the bat's wing or frog's foot under the microscope, the flow of blood in the small arteries, capillaries, and veins, may sometimes be seen at one and the same time, rapid and free in one part; slow and interrupted in another. Where the stream is slow and interrupted, we observe that the artery leading to the part is more or less constricted; where, on the contrary, the flow of blood is rapid and free, we observe that the artery leading to the part is wide and dilated. This difference in the rapidity of the circulation in two adjacent parts at one and the same time is owing to the circumstance that the heart's action necessarily operates with greater power on the blood in the dilated than in the constricted arteries.

The important fact that in a dilated artery the flow of blood is more rapid and free than in a constricted artery, was discovered independently by Mr. Paget*

---

* Lectures on Inflammation, at the Royal College of Surgeons, 1850.

and myself.*   I also showed and illustrated by delineations of appearances observed in the frog's web under
the microscope, how that when an artery is constricted
and dilated at different places, as it is sometimes found
to be, the stream, indeed, on entering a constricted from
a dilated part becomes accelerated, and on entering a
dilated from a constricted part becomes retarded in its
course.

The state of partial dilatation of an artery here
referred to as that on which the retardation of the
stream depends, it will be observed, is a physical condition quite different from the general dilatation of an
artery, which permits the heart's force to operate on the
stream with full effect.

Professor Marey, of Paris,† who has written with
great ingenuity on the subject, though he has not given
much attention to the observation of the microscopical
phenomena of the circulation in the extreme vessels,
erroneously attributes to Professor Claude Bernard the
discovery of the fact under notice.

Sometimes, as has been said, an artery may be seen
to become constricted even to closure, so that the flow
of blood in it is arrested.   It is worthy of remark that
when an artery under view is thus beginning to close in
the animal is agitated, and that soon its struggles put
an abrupt stop to the observation.   When the animal

* " Essay on Inflammation," in Guy's Hospital Reports for 1850.
This Essay, it may be mentioned, left my hands before Christmas,
1849.
   † " Physiologie Médicale de la Circulation du Sang, basé sur
l'Étude graphique des Mouvements du Cœur et du Pouls artériel."
Paris, 1863.

becomes quiet, so that its web can be again brought into the focus of the microscope, the artery may, perhaps, be found still constricted ; but it will be seen, in a short time, to dilate again with re-establishment of the stream of blood.

Constriction of an artery may be excited at pleasure by the application of cold, by making slight pressure over the vessels, or by touching the web with some chemical irritant.

When the constriction of an artery takes place, if the vessel has not quite closed in, the flow of blood, though retarded, may continue direct ; but it is sometimes seen to become retrograde by regurgitation from anastomosing vessels, and after continuing so for a more or less considerable distance, to join the direct stream in another anastomosing artery. If the constriction be so great that the artery is nearly or wholly closed in, the flow of blood, as just stated, ceases altogether at the place ; but into the artery below the constriction, a stream enters in a retrograde direction by one branch, and escapes in a direct course by another. From the artery above the constriction, the stream passes off by the first considerable branch.

I now come to notice the variations in the degree of fulness of the vessels of the rabbit's ear, as observed with the naked eye. In my paper on the rhythmical contractility of the veins of the bat's wing,* I refer, in a note, to the tonic constrictions of the artery in the ear of the rabbit, which I observed to take place in a

* Philosophical Transactions for 1852, p. 133.

manner similar to that in which the same phenomenon presents itself in the web of the frog's foot and in the web of the bat's wing. If we hold the ear of a rabbit, from which the fur has been shaved, against the light, and watch the artery which runs up the middle, we shall see it, at short but irregular intervals, become constricted, as manifested by the narrowing of the red column of blood which it contains. When this occurs, the ear becomes blanched and cold. The constriction of the artery now giving way to dilatation, as manifested by the increasing breadth of the column of blood which it contains, the whole ear becomes red from engorgement of its vessels with blood, and at the same time hot.

In short, we see that in proportion as the artery becomes constricted, the veins become small from the diminished afflux of blood, and in proportion as the artery becomes dilated the veins become large and distended with blood. Though, on account of the thickness and opacity of the skin, we cannot perceive the details of the process as we can do in the web of the frog or the wing of the bat under the microscope, there can be no doubt that the varying degree of the vascular fulness of the ear—the increased redness with heat or the diminished redness with lowering of the temperature—depends on the relaxation or contraction of the muscular walls of the artery occasioning dilatation or constriction of its calibre, exactly as we can so directly and distinctly observe in the web of the frog's foot or in the web of the bat's wing.

It is to be particularly kept in mind that there is *nothing rhythmical* in the alternate constrictions and dilatations of the artery of the rabbit's ear, any more than there is in the case of the constrictions and dilatations of the arteries in the frog's foot or the bat's wing. The constrictions and dilatations of the artery in the rabbit's ear are, however, seen to alternate more frequently than those of the arteries in the web of the frog or bat.

Such is an account of the real mechanism by which the rabbit's ear becomes injected with blood, both in the physiological state and after section of the sympathetic in the neck. Dr. Brown-Sequard, in describing the phenomenon of vascular injection after section of the sympathetic, speaks as if all the vessels—capillaries and veins, as well as arteries—were directly paralysed by the section, whereas they become gorged merely in consequence of the greater quantity of blood poured into them through the paralysed and dilated arteries. The capillaries do not become paralysed, the fact being that they have no muscular fibres in their walls to be paralysed. The veins do not become paralysed either, for though they have tonically contractile walls, there is reason to believe, as we shall see below, that their muscular fibres are not under the government of the same nerves which govern the contractions of the arteries.

The constrictions which the artery in the rabbit's ear is seen to undergo have been described as rhythmical, and the artery itself viewed as *an accessory arterial* *heart.** By its dilatation the artery, according to Dr. Schiff, receives and partly draws in the blood, while by its contraction it propels the blood onwards, and thus supports the circulation.

I have already said that the constrictions of the artery are *not rhythmical,* and that they are exactly of the same nature as those which the arteries in the web of the frog's foot or of the bat's wing may be observed under the microscope to present, though usually at longer intervals. To this it is now to be added, that the fibres of the muscular coat of the artery of the rabbit's ear present the same microscopical characters as those of the muscular coat of the arteries in the web of the frog's foot or of the bat's wing. Like the latter named arteries, also, the artery of the rabbit's ear is accompanied by a nerve in which, along with the sensitive nerves of the skin, there are contained vaso-motor fibrils.

What, let it be asked, is the effect on the circulation of the constriction of an artery in the web of the frog's foot or of the bat's wing which we see take place under the microscope? So far from being an acceleration of the flow of the blood in the capillaries and veins, it is a retardation if the constriction be only partial, and complete stagnation if the constriction be to closure of the lumen of the artery.

Microscopical observation of the phenomena in the frog's foot or bat's wing, however, is not necessary to

---

* " Ein accessorisches Arterienherz bei Kaninchen," von Dr. Schiff, in Frankfurt am Main. In Vierordt's " Archiv der physiologischen Heilkunde," Band xiii, 1854, pp. 523-7.

show the inaccuracy of Dr. Schiff. The phenomena observable in the rabbit's ear with the naked eye, if properly interpreted, are inconsistent with his view. In the first place, in the absence of a valve, the blood would be driven back as readily as forward by the constriction of the artery. But to say nothing of this, and to admit, for the sake of argument, that the blood is propelled onwards by the contraction of the artery, the capillaries and veins ought to become enlarged when the artery contracts, while when the artery dilates the veins ought to appear collapsed. But the reverse of this is the case. Watching the artery, we see it become larger and larger from the root to the tip of the ear, and, *pari passu*, we see the veins become distended. They then continue so as long as the artery remains dilated, but as soon as the artery becomes constricted the venous trunks become less turgid, in consequence of the diminution of the afflux of blood to them, and in consequence of the tonic contraction of their walls moving much of the blood distending them towards the heart. The blood in the capillaries and venous radicles, judging from what we see in the webs of the frog and bat under the microscope, no doubt becomes stagnant. After the death of the animal the veins are found still filled with blood.

The ear, which was red and hot during the turgid state of its vessels, becomes blanched and cold when the artery is constricted, in consequence of the diminished afflux of blood to it. This anæmic and cold stage, as we may call it, after continuing a short, though indefinite, time, is succeeded by the hyperæmic and hot stage, as before. By stroking gently the ear with a damp warm sponge, we may, at pleasure, in general bring on the dilatation of the artery and the injection of the vessels.*

From this account of the circulation in the rabbit's ear, it will be seen that in it we have a beautiful exemplification of how the supply of blood to a part is regulated by the tonic contractions of the arteries, independently of any variation in the force of the heart's action.

The contractile power of the walls of arteries, we have seen, is paralysed by section of the sympathetic in the neck, or destruction of the roots of that nerve in the spinal cord, and the vessels become much dilated. After a time I have found the walls of arteries in the frog's web, after section of the ischiatic, regain their contractility, so that the vessels were observed to become constricted occasionally. The constrictions, however, were of unregulated occurrence, in comparison to what they are in the natural state.

Notwithstanding that the tonic and non-rhythmical character of the contractions of the artery of the

---

* At the end of a journey, if the ears of a horse are cold, a careful groom, before putting it up in the stable, diligently strokes the ears through his hands until they get warm. By this treatment, it is believed that the horse is rendered less liable to suffer from the cold to which it has been exposed.

rabbit's ear is so obvious, whether as a matter of mere direct observation, or as one of observation and deliberation combined, Dr. Schiff's view has been accepted by distinguished physiologists, and the alleged rhythmical contractions of the artery of the rabbit's ear collated with the real rhythmical contractions—the systole and diastole—of the veins of the bat's wing.

Rudolf Virchow,* the celebrated Professor of Pathology in the University of Berlin, and the well known liberal member of the Prussian Parliament, considers the contractions of the artery of the rabbit's ear, as described by Dr. Schiff, to be " a proof that a real rhythmical movement does take place in the arterial walls ;" and says that the only counterpart to it " exists in the movements which had previously been observed by Wharton Jones in the veins of the wings of bats." He adds that he has studied these phenomena. I cannot, however, refrain from remarking that if Professor Virchow had observed and reflected on them a little more carefully, he would not have expressed himself on the subject in the loose manner he has done.

Professor Valentin, of Bern,† who has done so much good work in physiology, examines the subject somewhat in detail, and correctly describes the correlations of the phenomena of alternate vascular fulness

---

* "Cellular Pathology, as based upon Physiological and Pathological Histology." Translated by Frank Chance, B.A., M.B., Cantab. London, 1860, pp. 116, 117.

† "Versuch einer physiologischen Pathologie des Herzens und der Blutgefässæ." 1866.

and emptiness. He does not, however, refuse to admit that the constrictions of the artery have the effect of propelling the blood onwards in its course. His conclusions are :—" These facts show that an influence, as yet unknown in its details, which acts on the nerves, energetically excites contraction of the larger and smaller arteries of the rabbit's ear. It subsequently becomes exhausted, but after a time it recovers itself, and excites the contraction. This independent alternation was the reason why a *heart in the rabbit's ear* was spoken of. It is, however, to be admitted that here we have no exceptional action, but only a place which, on account of its transparency, permits us to observe a phenomenon which is probably widely spread. As women with delicate skin blush not only on the cheeks, but also on the chin, the forehead, or the ears, and as any one becomes pale over the whole face, so the rabbit's ear becomes now red and now pale, because in that part of the animal the most favourable conditions exist for the visibility of the alternations in the fulness of the blood vessels. We here at the same time directly recognise how the column of blood in the artery penetrates, and gradually makes way for itself by distending the vessel when the contraction of its walls no longer forbids."

Valentin gives, at p. 371, a figure of the vascular distribution in the rabbit's ear, which, together with his description, would lead us to suppose that small arterial ramifications open directly into venous trunks. Both the live rabbit and a drawing of the blood-vessels of the ear, from the life, made many years ago, before me show that this is a mistake similar to that which was once committed in regard to the arteries and veins of the bat's wing.*

Dr. Rolleston, the distinguished Professor of Anatomy and Physiology in the University of Oxford, in the Address he delivered at the thirty-sixth annual meeting of the British Medical Association at Oxford, in 1868, commenting on the constrictions and dilatations which alternate in the caliber of vessels, says that they take place, "as is well known, spontaneously, as the phrase goes—whether rhythmically or not, still chronometrically in relation to the needs of the animal and its tissues ; in the arteries of the rabbit's ear (Funke, 'Physiologie,' vol. ii., p. 771, citing Schiff and Callenfels); in the veins of the bat's wing (Wharton Jones, Phil. Trans., 1852) ; in the arteries of the frog's web (Lister, Phil. Trans., 1858, p. 653)."

If Dr. Rolleston will again examine for himself the phenomena, on which he thus discoursed *ex cathedrâ,*

---

* See the correction of this error in my Paper on the "Rhythmical Contractility of the veins of the Bat's Wing," p. 134. M. Marey (Op. cit., p. 350) quotes M. Sucquet ("De la Circulation du Sang dans les membres et dans la tête, chez l'homme." Paris, 1860), as "having discovered that in the limbs and head of the human body there is a vascular apparatus of a special disposition, the characteristic of which is, that certain arteries and veins communicate with each other by wide channels, so that the blood can pass freely from the arteries into the veins, thus forming a circuit of derivation by which too great fulness of the vessels is allowed to run off." It could only have been from injected preparations that M. Sucquet inferred the existence of such an arrangement. But injected preparations, I do not hesitate to assert, are of *no scientific value* in proving any one point in respect to the circulation in the extreme vessels.

he is, I am persuaded, too good an observer to remain
longer in doubt as to which contractions are rhythmical,
like the systole and diastole of the heart, and which
are merely tonic in their character ; and if he will
again read my essay on the state of the blood and the
bloodvessels in inflammation, in " Guy's Hospital Reports for 1850," he will find that he has overlooked the
detailed account which I give at p. 7, of the non-
rhythmical constrictions and dilatations of the arteries
of the frog's web. Of this Professor Lister was well
aware when he wrote eight years after. I am quite
sure, therefore, that if Dr. Rolleston will re-examine
Professor Lister's paper, he will find no expression in it
calculated to convey the idea that that gentleman gives
the observation attributed to him as his own. Moreover, if Dr. Rolleston will again read my paper on the
rhythmical contractility of the veins of the bat's wing, *
he will find at page 133 that I refer not only to the non-
rhythmical contractions and dilatations of the arteries
of the bat's wing as similar to, though not quite so
quick as those of the arteries of the frog's web, but also
(in a note) to those of the artery of the rabbit's ear—
phenomena which Dr. Schiff † two years after so
strangely misinterpreted.

* " Discovery that the Veins of the Bat's Wing (which are furnished with valves) are endowed with rhythmical contractility." In
the Philosophical Transactions for 1852.

† In Vierordt's Archiv, *ut supra.* Dr. Rolleston does not quote
from this, the original paper. A reference to it will show that
Dr. Schiff had studied my paper and repeated the observations on
the rhythmical contractions of the veins of the bat's wing.

NATURE OF ARTERIAL TENSION.

The blood propelled into the arteries by the systole of the heart distends them. During the diastole of the heart, the column of blood is compressed and forced on in its course by the tonic contraction, but chiefly by the elastic reaction of the stretched walls of the arteries. The intermittent flow imparted to the blood by the rhythmical or intermittent action of the heart, is thus at the same time converted, first into a remittent, and eventually into the continuous stream which it presents in the capillaries.

This tonic and elastic reaction of the arteries constitutes what is called *arterial tension.* It is no new or additional force, but is merely a conversion of that exerted by the heart to serve a purpose analogous to what is served by the compressed air in the reservoir of a fire-engine.

Arterial tension is not to be confounded, as is sometimes done, with the constriction of the small arteries, which we have seen exerts so important an influence in modifying the activity of the circulation in a part, independently of any variation in the force of the heart's action. Arterial tension, as just said, is mainly dependent on the *physical* property of elasticity of the walls of the large arteries, whereas constriction is wholly dependent on the *vital* property of muscular contractility, which the walls of the small arteries possess in so high a degree. Between arterial tension and constriction of the small arteries, indeed, there is a certain relation, for when any resistance to the free flow of blood in the extreme vessels occurs from constriction of the small arteries, the arterial tension is increased in consequence of the accumulation of blood in the larger arteries. When, on the contrary, the circulation is free in the extreme vessels in consequence of dilatation of the small arteries, the arterial tension is diminished.

THE CIRCULATION IN THE LUNGS CONSIDERED IN CONNECTION WITH THE SMALL DEGREE OF CONTRACTILITY WITH WHICH THE RAMIFICATIONS OF THE PULMONARY ARTERY ARE ENDOWED.

The number and size of the vessels and the closeness of the capillary network in the different parts of the body generally is in proportion to the structures to be nourished. In the lungs there is no such proportion. The arteries and veins are numerous and large, and the capillary network very close, though the amount of tissue to be nourished is small. This amount of tissue, such as it is, being cared for by the bronchial system of vessels, it may be said that the pulmonary system does not minister to the nourishment of any structures at all. In the close capillary network of the lungs the blood is sub-divided into minute streams, fitting it to undergo at once the process of aëration which consists in the interchange, by liquid diffusion, of the excess of carbonic acid gas which it holds in solution, for oxygen dissolved from the inspired air in the air-cells. In this little or no variation in the fulness of the stream and rapidity of the flow of blood seems to be required.

In the lung of the newt or frog I have not observed any material variation in the flow of the blood, independently of variation in the force of the heart's action, produced by constriction of the caliber of the arteries.

The pulmonary artery and its ramifications occupy an external stratum of the pulmonary cell, the pulmonary vein and its ramifications occupy an internal stratum, and the capillary network occupies the intermediate stratum. To the capillary network occupying the intermediate stratum the small arteries convey the blood, and from it the venous radicles proceed to the inner stratum to join the larger veins there.

The arterial and venous trunklets run at some distance from each other, as in the web of the frog. The last ramifications of the arteries become abruptly sub-divided to open into the close capillary network, while, by an equally abrupt union of the capillaries, the venous radicles are formed.

The pulmonary arteries have muscular walls, and, like the arteries of the frog's web, are accompanied by nerves, but it is to be observed that the muscular coat of the pulmonary arteries is not so thick as that of the arteries of the web. In the course of my observations I have never actually seen constriction of any of the arteries take place, as we so often see in the case of the arteries of the web of the frog's foot or bat's wing, except in a slight degree, and consequently I have not seen any material variation in the flow of blood occur from that cause independently of the action of the heart.

In the lungs, in short, the blood passes through the vessels in streams nearly as unvaried as the force of the heart's action by which they are propelled. The extreme facility and uniformity with which the blood in the healthy state, during the action of the air, glides through the capillaries, and the great sensitiveness of this free flow of blood in the capillaries of the lungs to imperfect aëration, are remarkable.

Morbid congestion, when it occurs in the proper pulmonary vascular system, commences in the capillaries, and is occasioned by imperfect aëration of the blood in its passage through these vessels.

In my " Observations on some Points in the Anatomy, Physiology, and Pathology of the Blood," first published as a report in the "British and Foreign Medical Review," in 1842, and afterwards in a separate form, I recorded, in illustration of this, the following fact bearing on the point: "The blood in the lungs of the frog was observed to be arrested in the vessels when the part of the lung under observation was acted on by carbonic acid gas directed against it in a small stream." And to this I added the remark that "the stoppage of the circulation in the capillaries of the lungs in asphyxia, it may be inferred, takes place in a similar manner."

Besides their pale red tint, the red corpuscles of the blood in its passage through the capillary network of the lungs, are remarkable for the appearance of semi-distension, softness, and flexibility which they present; in this respect resembling the red corpuscles in inflammatory blood; but differing from the red corpuscles of the blood in the capillaries of the web, which are flat and dark.

In the lung of a frog which had collapsed under examination, and in which the circulation had stopped, the red corpuscles were flat, collapsed, and dark-looking, like what they are in the web, and like what they were, though in a greater degree, after the action of carbonic acid gas in the experiment just mentioned.

The circulation in the lungs may, to a certain extent, be collated with that in the web of the frog's foot, after section of the ischiatic nerve, in respect to the rapidity of the flow and the fulness of the vessels with red blood, —lymph spaces next their walls being little evident, and, in consequence of the rapidity of the stream, accumulations of white corpuscles not so readily taking place.

THE RHYTHMICAL CONTRACTIONS OF THE VEINS OF THE BAT'S WING CONSIDERED AS AN AUXILIARY POWER IN THE CIRCULATION.

The demonstration by Malpighi* and Leuwenhoek † of the blood flowing from the small arteries through the capillaries into the rootlets of the veins; in the transparent parts of living animals—frogs and fishes— viewed under the microscope, was complemental of Harvey's great discovery. But, though the general doctrine of the circulation was thus fully established, a question arose as to the forces which propel the blood

* " Opera Posthuma," pp. 91, 92. Londini, 1697.
† " Opera Omnia." Tom. ii., p. 174, &c. See also, Haller, " Opera Minora." Tom. i., Lausannæ, 1763, pp. 205, 6, 7.

onward in its circuit, and has continued to be agitated down to the present time.

Physiologists have found it difficult to admit that the action of the heart is alone sufficient to maintain the circulation. As an auxiliary force they used especially to appeal to the action of the arteries, under the mistaken notion that the pulsations of these vessels are owing to rhythmical contractions of their walls, whereby, like a second heart, they propel the blood onwards. Physiologists are, however, now agreed that the arterial pulse is owing, not to rhythmical contractions,* but to widening and elongation of the already full arteries, caused by the forcible injection of more blood into them by the systole of the heart. The blood, in consequence of the resistance to its onward movement through the capillaries, not escaping from the arteries as quickly as it is thus injected into them, distends these vessels in the direction of both their width and length ; and it is this distension which is felt as the pulse by the finger, on being applied with a little pressure over an artery— the artery at the wrist for example. The distension of the arteries being dependent on the systole of the heart, the arterial pulse, except in the smaller arteries, in which it is a little later, is felt at the same moment as the beat of the heart.

The walls of the arteries having elastic and muscular tissues entering into their structure, are endowed with

---

* The exceptional misapprehension of the nature of the contractions of the artery of the rabbit's ear has been above commented on at pp. 81-7.

both the physical property of elasticity, and the vital property of tonic contractility. By their elasticity the arteries are brought back to their previous width and length when the force by which they have been distended has ceased to act. The contractility of the arteries, we have seen, is not rhythmical or alternating with relaxations like that of the heart, but is tonic or continuous in operation, though with variable force and occasional relaxations. The general effect of the vital contractility of the walls of the arteries is thus much the same as that of their elasticity. But mere elasticity could not have served the same purpose as that which is served by tonic contractility, seeing that it was necessary that the force constricting the arteries should vary according to circumstances, whereas the elastic force operates uniformly in proportion to the extent to which the elastic body has been previously made to deviate from its state of rest or equilibrium.

The arteries, as thus endowed, promote the circulation only by the pressure resulting from their elastic and tonic reaction on the column of blood propelled into them by the systole of the heart. Thus, during the diastole of the heart, the aorta and pulmonary artery with their ramifications are exerting the elastic and tonic reaction of their walls on the blood by which they are distended. This constitutes, as above explained, *arterial tension*. The blood is thereby forced on in its course (regurgitation being prevented by the semilunar valves), and the intermittent flow, imparted to it by the intermittent action of the heart, changed first into a remittent and eventually into a continuous stream, such as it presents in the capillaries.

Many physiologists, still considering the heart's action insufficient for the circulation, have sought in the capillaries for the additional power required. It has been conjectured either that some propulsive force is actually exerted by the capillaries, or that certain attractions and repulsions are in operation within them, whereby the blood is drawn from the arteries on the one hand and repelled into the veins on the other. For the conjecture that the capillaries exert a propulsive force, there are no grounds whatever. Observation shows that these vessels are merely passive tubes—that they do not even possess tonic contractility. An important modifying influence is indeed, as I have above shown, seen to be exerted on the flow of blood in the capillaries by the constricted or dilated state of the small arteries opening into them. Thus, when the arteries of a part are dilated, the flow is rapid and free, but when constricted, the flow is impeded, and congestion of blood corpuscles tends to take place by regurgitation into, or retention in, the capillaries to which the constricted arteries lead. But in neither case is the circulation generally influenced.

As to the conjecture that certain attractions and repulsions are in operation within the capillaries, whereby the circulation of the blood is promoted :—Microscopical observation of the web of the frog's foot and bat's wing shows, indeed, that in the small arteries, capillaries, and rootlets of the veins there are certain attractions and repulsions among the blood corpuscles, and also between them and the walls of the vessels, which influence somewhat the passage of the blood in the part, but do not promote the circulation generally.

Amidst the uncertainty which thus appears to have existed in regard to the forces moving the blood, the discovery of the rhythmical contractions of the veins of the bat's wing showed that when, from some peculiarity of structure, such as the very extended wing of the bat, an auxiliary to the heart's action is really required, such is called into play; but of a kind simple and effective in comparison with those hitherto conjectured.

The muscular coat of the veins of the bat's wing, at the same time that it differs in the rhythmical character of its contractions, differs correspondingly in respect to the microscopical characters of its component fibres from that of veins generally as well as from that of arteries. The tissue of the rhythmically contractile muscular coat of the veins of the bat's wing, examined under the microscope, is seen to consist of circularly disposed and closely coherent fibrils of a granular semi-transparent aspect, quite different from the clear and transparent-looking fibrils composing the muscular coat of the arteries. Notwithstanding this, it is erroneously stated in a popular Treatise on Anatomy that the tissue of the muscular coat of the veins of the bat's wing is similar to ordinary unstriped muscular tissue.

The rhythmical contractions and dilatations of the veins of the bat's wing are, in the natural state, con-

tinually going on ; but sometimes with greater, sometimes with less force, and sometimes with greater,
sometimes with less rapidity. The average number of
contractions in a minute I have found to be ten. In
contracting, the vein closes in on an average to about a
fourth of its whole width. This constriction takes place
slowly, but the dilatation which supervenes takes place
quickly and as if with a jerk.

During the contraction, the flow of blood in the
veins is accelerated. On the cessation of the contraction the flow is checked, and regurgitation of the
blood tending to take place, brings the valves into
play.

It is to be observed that in determining the flow of
blood in the veins, the action of the heart is still concerned, as well as the contractions of the veins themselves. Sometimes the heart's action is sufficient to
keep up a pretty steady flow in the veins, this being
only accelerated at each contraction of these vessels.

The observation which I, at the same time, made,
that the veins of the ears of the long-eared bat are
unfurnished with valves, and are not endowed with
rhythmical contractility, and that the onward flow of the
blood in them is uniform, indicates that under ordinary
conditions no force in addition to that of the heart
is necessary. In short, "it illustrates how that the
heart's action is sufficient of itself for the circulation of
the blood in the body generally; but that being sufficient for that purpose only, the supplementary force
of rhythmical contractility of veins, supported by the

presence of valves, is called forth to promote the flow
of blood in the wings, which, on account of their extent,
are, as regards the circulation, in a considerable degree,
though not entirely, beyond the sphere of the heart's
influence." To this it may in conclusion be added,
that had the heart been made powerful enough for the
circulation through the wings, it would have been, perhaps, too powerful for the rest of the body.

In a popular Manual of Physiology, it has been stated
that the veins of the rabbit's ear are endowed with
rhythmical contractility, on the alleged authority of Dr.
Schiff. In this an application is erroneously made to
the veins of what Dr. Schiff really wrote in respect to
the artery of the rabbit's ear.

I need scarcely repeat, after what I have above
written (p. 81-7), that neither the artery nor the veins
of the rabbit's ear are endowed with *rhythmical* contractility. I have shown that as regards microscopical
structure, the muscular coat of the artery of the rabbit's ear is similar to that of other arteries universally
admitted to be endowed with no other than tonic contractility. I have now to state that the muscular coat
of a vein of the rabbit's ear was observed under the
microscope to be similar in the character of its fibrils
to the muscular coat of other tonically contractile veins
of the same size, and to differ altogether from the
muscular coat of the rhythmically contractile veins of
the bat's wing. It is moreover to be remarked that
like the veins of the ear of the long-eared bat, which

are destitute of rhythmical contractility,* the veins of the rabbit's ear have no valves, as was shown by the manner in which the blood remaining in them after death admitted of being pressed back from trunk to branches and thence towards the capillaries.

In the influence which the stream of lymph propelled into the caudal vein from the caudal heart of the eel exerts on the flow of blood on that vessel, we have another, though incidental and partial, example of a force auxiliary to that of the heart.†

---

* Phil. Trans., *ut supra*, p. 135

† The Caudal Heart of the Eel, a lymphatic heart, &c. In the " Philosophical Transactions " for 1868, pp. 675—683.

I have observed under the microscope an analogous influence exerted by the anterior lymph hearts of the frog.** It is into a vein at the lower and posterior border of the heart that the lymph is propelled and not into the jugular, as supposed by the late Professor Johannes Müller, the discoverer of the lymph hearts of the amphibia. From its close proximity to, and connexion with, the lymph heart, the jugular vein is, at each systole, dragged towards it, and at each diastole recoils from it by elasticity. This rhythmical backward and forward movement of the vessel longways appears to have been mistaken by Professor Müller for alternate turgidity and collapse, occasioned by the intermittent propulsion of lymph into the vein from the heart.

** Proceedings of the Royal Society, 1868.

CHAPTER X.

FROM the account of the phenomena of the circulation in the rabbit's ear in the last chapter, so far as they are observable by the naked eye, it was seen that we have there an exemplification of how the supply of blood to a part is regulated, independently of any variation in the force of the heart's action, by the constriction of the calibre of the small arteries produced by tonic contraction of their walls. And, at the same time, an exemplification of how the temperature of the part rises with the increased afflux of blood to it, through the dilated arteries, and falls when, by the constriction of the arteries, the supply of blood is diminished. Along with the increased afflux of blood also, it is to be observed that the vital energies of the part are exalted; while with a diminished afflux of blood they are lowered.

The contractions and relaxations of the muscular walls of the small arteries on which the variations in their width referred to depends are, we have seen, under the government and control of the vaso-motor nerves. That these are derived from the sympathetic, as before stated, is demonstrated by the following investigation.

Pourfour du Petit in 1727,* published the observation which he first made in 1712, that in the dog, division of the sympathetic nerve in the neck has for result the following phenomena :—half closure of the eyelids ; partial projection of the third eyelid over the cornea ; sunk appearance of the eyeball ; vascular redness and swelling of the conjunctiva with a collection of mucus at the inner corner; contracted state of the pupil. These effects, Petit attributed to the division of the sympathetic and not to that of the vagus, which is unavoidably divided at the same time on account of its being so closely connected with the sympathetic in one sheath.

That the eyes are thus affected by division of the sympathetic nerves in the neck, the more recent experiments of Reid, Claude Bernard, Brown-Sequard, and many others, have confirmed.

Another result of division of the sympathetic in the neck, Claude Bernard discovered, in experimenting on a rabbit, to be great redness and heat of the ears and side of the head. This was shown by Brown-Sequard to be owing to the increased afflux of blood to the parts, from the dilatation of the vessels. It is to be observed that the state of redness and heat of the ear

---

* Mémoires de l'Académie Royale des Sciences, pour 1727, p. 1. Amsterdam, 1733.

of the rabbit, above described at p. 80 as alternating with the opposite condition of paleness and coldness, exemplifies this state of redness and heat which is induced for a continuance by section of the sympathetic in the neck.

When, then, the sympathetic in the neck of a dog, cat, or rabbit is cut, vascular fulness in the corresponding side of the head ensues,—the ear becomes dark red, the conjunctiva and nasal mucous membranes turgidly injected, while, accompanying and dependent on this vascular fulness, there is increased heat of the parts. The same effects on the blood vessels of the parts named, and also contraction of the pupil, presenting themselves when the spinal cord was experimented on between the sixth cervical and ninth or tenth dorsal vertebra, according to Brown-Sequard, it is justly concluded that the region of the spinal cord comprehended between these limits is. the spinal centre of the sympathetic fibrils under notice.

Professor Valentin was the first who suggested that the iridal sympathetic has roots in the cervical portion of the spinal cord.*

According to the experiments of Budge, the communicating branch between the hypoglossal nerve and superior cervical ganglion, conveys from the medulla oblongata to that part of the sympathetic which governs the contractions of the radiating muscular fibres of the iris, a set of motor fibrils in addition to those which the sympathetic in the neck, below the superior cervical

* De Functionibus Nervorum, pp. 109—114. Bern, 1839.

ganglion, down to the inferior, receives. The part of
the medulla oblongata indicated is, therefore, viewed by
Budge as a *superior centre of the iridal sympathetic
nerve*, in contradistinction to that part of the spinal
cord between the sixth cervical and fourth dorsal
vertebra, which he calls the *inferior spinal centre of
the iridal sympathetic.*

It was shown, in 1846 by Dr. Biffi that when the
pupil had become contracted from paralysis of the
radiating fibres of the iris, after section of the sympathetic in the neck, irritation of the nerve above the
section excited dilatation of the pupil. In like manner
it occurred to Dr. Brown-Sequard,* that irritation of
the sympathetic would cause constriction of the blood
vessels of the head, which had become dilated from
paralysis of their walls, in consequence of section of
the sympathetic in the neck. Accordingly, on per-
forming the experiment of galvanising the sympathetic
in the neck above the place of section, Dr. Brown-
Sequard found that the increased redness and heat
which had supervened on the section of the nerve, were
for the time diminished, in consequence of the contraction of the vessels which are under the influence of the
sympathetic. This, it is to be observed, was the induc-
tion of a state of paleness and coldness exemplifying
that which naturally alternates with the opposite state
of redness and heat in the rabbit's ear, as above described at pp. 80-1.

* Sur les Résultats de la Section et de la Galvanisation du Nerf
grand sympathetique au Cou. Mémoire lu à l'Académie des Sciences,
le 16 Janvier, 1854. In the Gazette Médicale de Paris, année 1854.

The occurrence of vascular injection of the eye after
section of the sympathetic in the neck, I attributed (in
the first edition of my work on " Ophthalmic Medicine
and Surgery," published in 1847) to paralysis of the
walls of the blood vessels of the eye ; and in my essay
" On the State of the Blood and the Blood Vessels in
Inflammation," published in Guy's Hospital Reports in
1850, an effect of section of the ischiatic nerve in the
frog is stated to be dilatation of the arteries, and a
fuller and more rapid circulation of the blood in the
web.   The following experiment was given in exemplification :—The ischiatic nerve of the left leg of a frog
being divided, the arteries of the web were found, on
examination under the microscope, to be dilated, and
the stream of blood in them fuller and more rapid.
The blood in the capillaries and veins, especially,
appeared to be unusually loaded with red corpuscles.
The relaxed arteries yielding more readily to the blood
injected into them at each stroke of the heart, their
pulsations were more evident.   The general effect to
the naked eye was increased redness, not only of the
web, but of the whole limb below the place of section.
Four days after the section the arteries were found still
dilated, and the circulation in the web very free.   The
epidermis was exfoliated.   On the right or uninjured
side, the arteries and circulation in the web remained
unaffected, and the epidermis was not exfoliated.

It is to be observed that in this experiment the
relaxation of the muscular coat of the arteries, on which
their dilatation depended, was the effect of section of the

sympathetic or vaso-motor fibrils bound up in the ischiatic nerve, and not of the section of the proper spinal fibrils of that nerve.

Though after section of the ischiatic nerve in the experiment mentioned, the arteries did not lose their contractile power, their tendency to become constricted was irregular in comparison with that manifested by the arteries of the opposite extremity, in which the ischiatic nerve was not cut. There being no longer any communication with the reflex centre, there was no longer any co-ordination of the actions of the vessels.

The increased redness and heat of the ear and side of the head in the experiment of cutting the sympathetic in the neck of a cat or rabbit, for example, are, we have seen, obviously owing to the increased afflux of blood to the parts, permitted by the dilatation of the artery, as we have seen to be the case in the natural state at short intervals.

That this increased afflux of blood is fundamentally owing to a fuller and more rapid flow in the arteries in consequence of their calibre having become dilated from impaired contractility of their walls, was to be inferred from my microscopical observations and experiments on the frog's web just quoted. There is no change in the capillaries and venous radicles, except in so far that they are more filled in consequence of the increased quantity of blood poured into them through the dilated arteries. The veins are so filled to their very walls that the lymph spaces no longer exist.

The similar paralysis of the muscular walls of the

arteries produced by damage to the vaso-motor nerves at their roots in the spinal cord, with a corresponding disturbance of the circulation in the part implicated as the result, was shown by the following experiment on the frog :—

The lower part of the spine of a frog was crushed by means of a pair of pliers, whereupon the posterior extremities and their webs were seen with the naked eye to have become red from vascular injection. The redness was most marked on the inside of the legs, where the skin is pale and delicate. The posterior extremities were paralysed immediately after the injury to the spine, but when examined a few hours after, the animal had so far recovered as to be able to move its hind legs in crawling like a toad. It continued, however, quite unable to leap.

Under the microscope the veins and capillaries were seen to be gorged to their very walls with blood much loaded with red corpuscles. The arteries were rather dilated, and the blood flowed freely and rapidly through them. No lymph spaces were to be seen, nor colourless corpuscles accumulated on the walls of any of the vessels. Here and there, indeed, small veins and capillaries were observed, in which the blood flowed sluggishly and in small quantity. At one place, there was slight extravasation and stagnation of blood corpuscles in the capillaries.

Altogether the circulation in the web was free, but not so very much so as we have seen it to be when the ischiatic nerve, with the vaso-motor fibrils of the sympathetic therein contained, has been cut. This was well shown by dividing the ischiatic nerve high up in the limb on one side in the same frog which had been previously subjected to the injury of the spine. In the frog referred to, both posterior extremities and their webs were so much congested that, as I have stated, they appeared red to the naked eye. On cutting the ischiatic nerve on one side, the corresponding limb became very much redder than that of the other side, and under the microscope the arteries were seen to be wider, the flow of blood in them fuller, and the veins and capillaries proportionally more gorged with blood laden with red corpuscles. No lymphatic spaces were to be seen.

CHAPTER XI.

EFFECTS OF IMPLICATION OF THE VASO-MOTOR NERVES IN INJURY OR DISEASE ON THE TEMPERATURE, NUTRITION AND VITAL ENERGY OF PARTS.

EFFECT OF IMPLICATION OF THE VASO-MOTOR NERVES IN INJURY OR DISEASE ON THE TEMPERATURE OF A PART.

THE generation of heat in the nutritive processes of the body does not come under discussion here. We start with the heat already generated and of which the blood is the vehicle for distribution to the different parts of the body, and have to inquire merely into the local variations of temperature determined by variation in the afflux of arterial blood to a part.

Nature of the sensations of temperature.—Heat in its various degrees from cold to burning is a sensation peculiar to the sense of touch, just as the sensation of light in its various degrees of brightness and colour is the sensation peculiar to the sense of sight. As the external agent *light* is the proper excitant of the optic nerve, so the agent *caloric* is a proper excitant of the tactile nerves. But it is to be remarked that the agent *light* has its source external to the body only, and thus impresses the retina from without, whereas the *caloric* which acts on the tactile nerves has a source within the body, namely, in the blood, as well as from without the body, namely, in the surrounding medium. The sensation of heat derived from the impression of the warm blood on the tactile nerves, therefore, is fundamentally as much an objective sensation as that excited by a warm body, such as warm air, in contact with the skin.

But as the sensations of light and colour may be excited by the impression on the optic nerve of other agents besides that of *light*, so the sensations of temperature may be excited by the impression or operation on the tactile nerves of other agents besides that of *caloric*. The sensation of heat or cold, thus arising, is that which is, properly speaking, subjective. Subjective sensations of temperature are, therefore, in their nature, morbid, and may be felt in opposition to the coincident operation of objective heat or cold. Sensations of temperature of this character are, accordingly, to be received as symptoms of some morbid state of the tactile nerves of the part or of their roots, or of the part of the nervous centre whence their roots spring.

By this, it may be explained how it is that in cases of paralysis, the patient may complain of sensations of heat or cold not in accordance with the objective temperature of the part as ascertained by the thermometer. As with lost sensibility of the retina and optic nerve, there is darkness before the eyes in the light, so with lost sensibility of the skin, there is a feeling of cold or absence of the sensation of heat, though the air bc warm and though the objective heat of the skin may not be lost.

Local variations of temperature as signifying increased or diminished afflux of blood to the part.— Increased afflux of arterial blood to a part is owing to dilatation of its arteries, and brings with it an increase of temperature. Diminished afflux of arterial blood, again, is owing to constriction of the arteries of the part, and is attended by a corresponding diminution in the supply of heat. Dilatation of the arteries implies . *diminished* action, and constriction of the arteries *increased* action of the vaso-motor nerves.

Increased heat of a part, therefore, is a manifestation of diminished action of the nerves governing the contractions of the muscular walls of its arteries ; while diminished heat or coldness of a part, on the other hand, is a manifestation of increased action of the same nerves.

Such local variations of temperature, within certain limits as regards degree and continuance, occur, we have seen, p. 80, in the healthy state.

A continued objective elevation of the temperature of a part above, or a continued objective depression of the temperature below, the natural standard is morbid, and among other causes, may depend on disease of the spinal cord implicating the roots of the vaso-motor nerves of the part affected. Thus, the elevation of temperature above the natural standard indicates that

the disease of the spinal cord is causing paralysis of the
vaso-motor nerves, while the depression of temperature
below the natural standard indicates that disease of the
spinal cord or its membranes is causing irritation and
excitement of the vaso-motor nerves.

*Influence of external temperature on the tempera-
ture of a part of the body.*—In a low surrounding temperature our hands, for example, feel cold from two
causes :—1st. The actual abstraction of heat by the
cold air.   2nd. The diminished afflux of arterial blood
arising from constriction of the arteries excited by the
external cold.

Under the circumstances described, the fall in the
temperature of the part is objective, but as the energy
of the tactile nerves is, at the same time, impaired in
consequence of the diminished supply of arterial blood
to the part, the sensation of cold may be greater than
what is directly dependent on the objective fall of temperature.   After a time reaction perhaps sets in and
our hands come to be warm—glowing warm—notwithstanding actual abstraction of heat by the cold air may
still be going on.   This is owing to constriction giving
way to relaxation of the walls of the arteries of the part
and consequent dilatation of their calibre, the effect of
which is increased afflux of arterial blood bringing with
it, as we have seen, an increased supply of heat—a
supply which is not only sufficient to make up for that
which is lost by abstraction by the cold air, but also to
maintain the warmth of the part.

In a high surrounding temperature we feel warm

from two causes: 1st, the warmth of the surrounding medium, which, if it is not high enough actually to
impart heat, abstracts little ; 2nd, the relaxation of the
walls of the arteries promoted by it and consequent
dilatation of their caliber, the effect of which is an increased afflux of arterial blood. As with diminished
afflux of arterial blood to the hand from constriction
of the arteries, there is numbness and diminished sensibility to heat, so, on the contrary, with increased afflux
of arterial blood to the part from dilatation of the
arteries, there is exalted sensibility to heat ; hence,
perhaps, in the one case the part feels colder than
it is, while in the other, the part feels the contact
of a warm body warmer than it really is, as in inflammation.

Thus it is that in anæsthesia there is subjective coldness, and in hyperæsthesia, subjective heat.

EFFECT OF IMPLICATION OF THE VASO-MOTOR
NERVES IN INJURY OR DISEASE ON THE
NUTRITION OF PARTS.

That along with the disturbance of the circulation
in a part, the walls of the arteries of which have been
paralysed by division, in some part of their course, of
the vaso-motor nerves distributed to them, or by
damage of the roots of the same nerves from disease
or injury of the spinal cord, there is deteriorated
nutrition, was shown in the experiments on the frogs
before referred to. Thus, on the side on which the
ischiatic nerve was cut, the epidermis was found four

days after to have exfoliated, whereas on the uninjured side nothing of the kind had occurred. Besides
exfoliation of the epidermis, extravasations of blood,
œdema of the leg and foot, and a tendency to ulceration and sloughing of the toes and webs, were also
sometimes observed after section of the ischiatic nerve.

After section of the ischiatic nerve in a dog the hair
has been found to fall off, leaving large bald patches,
and section of the fifth nerve in rabbits has been
followed by falling out of the whisker hairs of the
corresponding side of the face.   The hoof in horses
has been found to grow thick and irregular after
section of the nerve of the limb.

Dupuy, Dupuytren, and Breschet (" Journal de Médecine," tome 37) observed in the horse, after extirpation
of the superior cervical ganglion of the sympathetic
nerve, violent inflammation of the eye on the same
side.   Magendie and Mayer performed similar experiments, and found inflammation of the eye a constant
result of ligature of the sympathetic nerve in the neck.

Similar effects of injury, either to portions of the
nervous centres or to the nerves of the limb, have
been observed in man.   Thus, Sir Benjamin Brodie,
in his " Lectures on Pathology and Surgery " (p. 309,
1846), observes that " in a case in which the spinal
cord is injured in the middle of the back, you may
find, almost before you suspect that there is anything
wrong, a great slough over the sacrum—nay, pressure
of the mattress on the ankles will, in such cases, produce the same mischief.   I have known," he continues,

" mortification begin in the ankle within twenty-four hours after an injury of the spine."

When, in man, the nerve of a limb has been accidentally divided, the hand or foot remains disposed to become gorged with blood, and is liable to chilblains, ulcers, &c.

Professor von Walther * has recorded the case of a young woman aged 22, in whom inflammation of the right eye, ending in abscess and atrophy of the organ, supervened one month and a half after ligature of the common carotid near its origin from the arteria innominata. The operation was performed on account of a large aneurism of the right carotid, which had been caused by the patient slipping and falling backwards while carrying a heavy basket on her head. The size of the aneurism rendered the operation difficult, and Walther believed that the sympathetic nerve was injured. Although this case is adduced as an example of inflammation of the eye supervening on the accidental division of the sympathetic in the neck in man, it cannot be admitted as a pure example, for the general reaction and disturbance of the system, the cerebral congestion, the hæmorrhage, &c., which followed the operation, may have had a large share in bringing on the inflammation by which the eyeball was destroyed.

In the production of these and analogous morbid states resulting from the injury or section of nerves, much depends, it has been found, on the co-operation

---

* " Ueber die Amaurose nach Superciliar-Verletzungen." In Græfe und Walther's Journal, Band xxix., 1840.

of external circumstances, such as mechanical pressure, extremes of external temperature, &c., the injurious influence of which the affected parts appear no longer able to resist as they were in the healthy state.

According to Professor Rudolf Virchow, hyperæmia or congestion exercises no direct influence on the nutrition of the tissues. " If a special proof were still required," Professor Virchow goes on to say, " in order to complete the refutation of the assumption that hyperæmia exerts any influence on nutrition, which, in an anatomical point of view, is utterly untenable, we find a most apposite argument in the experiment of the section of the sympathetic." The resulting injection of the blood vessels, Professor Virchow continues, " may persist for days, weeks, or months, without the least appreciable nutritive disturbance necessarily arising therefrom ; the parts, although gorged with blood, are yet—as far at least as we are at present able to judge of this—in the same state of nutrition as before."

From this quotation, Professor Virchow would seem to have looked especially for manifestations of nutritive disturbance such as present themselves in common inflammatory congestion. That such is not now the view really entertained by Professor Virchow, however (nor indeed the view originally entertained by him), would appear from a note towards the end of his book communicated in MS. to the translator, in which the Professor admits that the section and paralysis of nerves do certainly exercise some influence upon the nutrition of the tissues, although, perhaps, only an indirect one. Professor Virchow appears, in fact, to have returned to a previous view published in 1854 in his "Manual of Special Pathology," in which he classes together the states arising from the section and paralysis of nerves under the name of "Neurotic Atrophy."

This kind of atrophy seems to be owing to a morbid increase of the activity of absorption in consequence of the increased rapidity of the circulation in the part which is a direct effect of the dilatation of the arteries. Atrophy, which is the result of general collapse of the arteries of a part, is, on the other hand, owing not to increased activity of absorption, but to a diminishing supply of blood. This is strikingly exemplified in the rapid shrinking of the tail of the tadpole which takes place as soon as the forelegs make their appearance, and the blood is diverted from the tail into new channels. This is the period of the metamorphosis into the *frog state*, when the branchial respiration is superseded by the development of the pulmonary respiration. It is not until the forelegs have become free that the young frog could be fitted to make its way out of the water to breathe air.

The state of constriction of the arteries from spasmodic contraction of their muscular walls, causing a diminished afflux of blood to the part with corresponding impairment of the heat, nutrition and vitality, is induced when disease or injury of the spinal cord or its membranes, instead of involving the roots of the vaso-motor nerves in destruction, is sufficient only to excite irritation of them.

The differences in the effect on nutrition, of an increased and a diminished afflux of blood to a part, was well illustrated in a case of transplantation of a flap of skin from the ball of the thumb to form a new eyelid, according to the Italian or Taliacotian method, which I have recorded.* The flap of skin, as long as its connection with the hand was maintained, was the seat of a considerable afflux of blood, and its edges in the line of union with the neighbouring skin of its new locality presented here and there florid and exuberant granulations, but after its severance from the hand, when all the supply of blood it received was merely that through its new vascular connection, the florid and exuberant granulations at the margin of the flap all disappeared, and subsequently the transplanted flap of skin itself shrank very considerably in size.

In limbs paralysed from injury of the spine, nutritive repair in some cases has gone on well, but in other cases the contrary has been observed. Thus, in the case quoted by Mr. Erichsen (Eve's " Surgical Cases," p. 90 ; and "New York Journal of Medicine," 1853. By U. D. Purple, M.D., of Greene, New York), of long persistent paralysis after a blow on the spine, in which the patient underwent the amputation of both thighs without feeling the slightest pain, so complete was the loss of sensation, healing of the stumps took place readily, and no unfavourable symptoms occurred in the progress of perfect union by the first intention.

---

* " Ophthalmic Medicine and Surgery." 3rd edition, 1865, p. 634.

Mr. Travers * mentions a case of an opposite kind. Besides fracture of the lumbar vertebræ, the patient suffered fracture of the humerus and fracture of the tibia. In consequence of the spinal part of the injury, paraplegia was produced, and it was found that the tibia did not unite, while the humerus did, showing that the paralysis was unfavourable to nutritive repair.

An explanation of the difference as regards nutritive repair in these two cases may perhaps be found in the circumstance that in the first case the vaso-motor nerves were wholly paralysed, and the arteries supplied by them being consequently dilated, the circulation was free, whereas in the second case the vaso-motor nerves were only irritated, and the arteries supplied by them being consequently constricted, the circulation was impeded.

In a case of incomplete paralysis from slowly increasing compression of the spinal cord, Ollivier found that there was coldness of the skin and exfoliation of dry epidermic scales. This was no doubt owing to constriction of the arteries from irritation of the vaso-motor nerves.

EFFECT OF IMPLICATION OF THE VASO-MOTOR NERVES IN INJURY OR DISEASE, ON THE VITAL ENERGY OF A PART.

An immediately and directly necessary condition for the maintenance of the vital energy of a part is the due circulation of blood in it. A diminished afflux of

* " Further Inquiry concerning Constitutional Irritation," p. 436.

arterial blood to the skin, at the same time that it is attended by coldness of the part, is a cause of impairment of the sensibility of touch.   In like manner, transient failure of sight may be occasioned by transient constriction of the arteries of the optic nervous apparatus arresting the circulation therein.   In already existing morbid states of the optic nervous apparatus, such an arrestment of the circulation is more prone to take place, and the resulting failure of sight is pronounced in a still greater degree.

Too great an afflux of blood operates in a manner not less injurious to the vital energy of a part.

Having thus studied the effects of implication of the vaso-motor nerves in injury or disease on the temperature, nutrition and vital energy of parts, the subjects which still claim attention before finally proceeding with our special investigation, are, first, the state of the blood and the blood vessels in inflammation and the nature of the inflammatory process ; and secondly, embolism or obstruction of the circulation in a part by clots of fibrin blocking up vessels.

CHAPTER XII.

STATE OF THE BLOOD AND THE BLOOD VESSELS IN INFLAMMATION, AND THE NATURE OF THE INFLAMMATORY PROCESS.

INTRODUCTORY OBSERVATIONS.

IT has not unfrequently been objected to the results obtained from microscopical observations on the web of the frog in the inflamed state, that we cannot safely argue from them as to the nature of the inflammatory process in man.

This, however, like many other general objections, is well or ill-founded, according to the sense in which the terms are employed. If in the objector's mind the word "inflammation" conjures up the idea of a pleurisy or pneumonia, of a meningitis or myelitis; nay, even of a conjunctivitis or iritis, with all the attendant symptoms, subjective as well as objective, functional disturbance and terminations, I admit that the objection is well founded. But if, on the contrary, all that we venture to deduce from our microscopical observations on the web of the frog in the inflamed state, be merely something concerning the general nature of the inflammatory process, something of the state of the blood and the blood-vessels in the inflamed part; something, in short, in elucidation of the question as to the *proximate cause* of inflammation, so much agitated by the pathologists of the last century, then I apprehend that the objection is ill-founded.

Still, there can be no doubt that even as regards the simple elucidation of the state of the blood and the blood-vessels in an inflamed part of the human body, microscopical observations made on a transparent part of the warm-blooded mammal would be more satisfactory than similar observations on the cold-blooded reptile, especially when we take into consideration the marked peculiarity of character which the red blood-corpuscles of the mammifera present.

Accordingly, in pursuing my researches into the nature of the inflammatory process, I have not failed to direct my attention to the microscopical study of the effects of wounds on the state of the blood and the blood-vessels in the web of the bat's wing; the bat being the only mammiferous animal which presents an external part of the body thin and transparent enough for examination under high microscopical powers. An objective $\frac{1}{8}$th of an inch, I have employed, with perfect facility. The long-eared bat, or the common gray house bat, I have found so tame as to lie quiet, with its wing under the microscope. Besides this advantage, the web of the wings of these bats is very thin, and contains little black pigment.

MICROSCOPICAL EXAMINATION OF THE BLOOD
AFTER BEING DRAWN FROM THE BODY, AND
ALSO WHILE WITHIN THE VESSELS OF THE
LIVING ANIMAL.

Human Blood.—A drop of blood drawn from a person in health, received on a glass slide, and spread out by being covered with a thin scale of glass, was forthwith examined under an eighth of an inch objective.

In the unpublished delineation of the appearances which was made at the moment, and which I have now before me, the corpuscles, both red and colourless, are seen dispersed confusedly about in the plasma.

The delineation (also unpublished) of the same drop of blood, as examined in about two minutes after, represents the following appearances:—The red corpuscles have become aggregated together like coins in rolls, and the fibrin of the plasma has coagulated into granules and fine pale fibres shooting in every direction on the glass slide. The colourless corpuscles, few in number, are seen scattered about, free from any connexion with the red corpuscles.

The transition from the state of the drop of blood as just drawn from the body, to that which it presented in about two minutes after, as represented in the two delineations before me, was as follows:—The red corpuscles first overlap each other in linear series, then, rising up on edge, become fully applied face to face. The rolls of corpuscles thus formed are disposed in the manner of an irregular network. The interspaces representing the meshes, are filled with plasma, in which colourless corpuscles, in a collapsed state, are seen free here and there.*

The red corpuscles having thus become aggregated in rolls, and the fibrin of the plasma coagulated and deposited on the glass slide in the shape of minute granules and fine pale fibrils, like acicular crystals shooting in a stellate manner, serum only remains in the meshes of the network of rolls, and the red corpuscles now no longer cohere with the same degree of force as they did at first. The rolls are, therefore, readily broken up by a slight touch on the superjacent scale of glass with the point of a needle. In consequence of the agitation thereby occasioned, currents of serum and disaggregated corpuscles take place. In a delineation (unpublished) which I have before me of matters in this stage of the observation, red corpuscles are seen glued to the glass slide by coagulated fibrin. Many being adherent by a single point of their surface are seen drawn out into a pear shape, under the pressure caused by the stream of serum and free corpuscles bearing on them.

The arrangement of rolls in a *network* is exhibited only when the blood is thinly spread out. When the blood is examined under the microscope in the form of a drop, the rolls are observed to be disposed in every direction, and thus constitute a *sponge work*, in the

---

* In my papers on " the Blood Corpuscle, in its different phases of development," in the Philosophical Transactions for 1846, I have described the *Amœba-like* changes of form which the colourless corpuscles of the blood of man and various animals undergo.

interstices of which the plasma or liquor sanguinis is contained.

The tendency of the rolls of red corpuscles to undergo disaggregation as soon as the fibrin has been deposited from the plasma, would show that their aggregation at the first is dependent in some manner on the fibrin. And this is confirmed by the well-known fact that in blood drawn from a person labouring under some acute inflammation, such as rheumatism, in which, with diminution of the number of red corpuscles, there is an increase in the quantity of fibrin in the plasma, the red corpuscles are found to become aggregated together both more rapidly and more closely.*

Blood of the Bat.—The blood corpuscles of the bat, both red and colourless, resemble those of the blood of man in shape and structure, but are somewhat smaller. The red corpuscles, $\frac{1}{4200}$th of an inch in diameter, are seen under the microscope to agglomerate into rolls, like

---

* The subject of the aggregation of the red corpuscles both in healthy and inflammatory blood, I have treated of in detail in the following writings :—

1. " Observations on some points in the Anatomy, Physiology, and Pathology of the Blood," in No. XXVIII. of the British and Foreign Medical Review, 1842.

2. " Report on the Changes in the Blood in Inflammation, &c.," in No. XXXV. of the British and Foreign Medical Review, 1844.

3. " On the State of the Blood and the Blood-vessels in Inflammation, &c.," chapter viii. in Guy's Hospital Reports, vol. vii., part i., 1850.

4. " On the Distinction between Healthy and Buffy Blood in minute quantities." In the Edinburgh Medical and Surgical Journal, vol. lx., pp. 309-311, 1843. In this paper, delineations of the close aggregation of the red corpuscles in buffy blood are given ; the first delineations of the kind, I believe, that ever appeared.

those of human blood, not only in blood out of the body, but also in blood within the vessels, under the conditions to be mentioned below. The colourless corpuscles occur in the same forms as in human blood. Measured while stagnant within the vessels of the living animal, the colourless corpuscles were full $\frac{1}{3000}$th of an inch in diameter.

As to the manner in which the blood corpuscles comport themselves within the vessels of the web of the bat's wing :—Within arteries and veins, the red corpuscles keep together in the axis of the stream, and do not manifest any tendency to adhere to the walls of the vessels, like the colourless corpuscles which occupy the spaces next the walls, where plasma, unmixed with red corpuscles, flows less rapidly, and which are named *lymph spaces.* When there is an impediment to the onward flow of blood, the red corpuscles agglomerate together in rolls similar to what we see in blood removed from the body ; but when the blood is again permitted to flow freely, the rolls are broken up, and the corpuscles are carried along confusedly mixed together. In the stream within the capillaries, the red corpuscles may often be seen arranged in a linear series, overlapping each other's edges ; but on the occurrence of any impediment to the flow of blood, the red corpuscles become aggregated, so that we may sometimes see a single long roll occupying the axis of a capillary, as represented in a drawing (unpublished) made from nature, now before me.

Colourless corpuscles may be seen rolling or sliding

sluggishly along the walls of the vessels—in arteries, occasionally, as well as in veins. They are sometimes adherent to the walls of the vessels. A colourless corpuscle thus adhering by a point, may now and then be seen drawn out into a pear shape by the force of the stream of red corpuscles, before becoming detached.

I never observed the escape of a corpuscle—red or colourless—through any pore of the wall of an unruptured capillary. A nucleus in the wall of a capillary is occasionally seen, presenting the appearance of a colourless corpuscle in the act of escaping through a pore; but on directing my attention to the point, and steadily watching, I have always recognised the true nature of the appearance.

Blood of the Frog.—The red corpuscles of the blood of the frog differ, as is well known, from those of the blood of man and the mammifera both in shape and structure. In the early embryo of man and the mammifera, the red blood-corpuscles are nucleated cells similar to those of the frog and other oviparous vertebrata; but it is an error to suppose that the smaller and unnucleated red corpuscle of the more advanced embryo, of the fœtus, and of the mammal after birth, is homologous or of the same nature with the large red nucleated blood-corpuscle of the early embryo. It corresponds, as I have shown, to the nucleus only.

If a drop of blood just drawn from a frog be examined under the microscope, the red corpuscles are seen to become, especially if the blood be that of a young healthy animal, somewhat aggregated together, lying flatways, and overlapping each other. A few may be observed here and there, raised up on edge, with their flat surfaces applied to each other.*

Within the vessels, when there is an obstruction to the free flow of blood, this aggregation of the red corpuscles of the frog, surface to surface, may be seen in a greater degree than takes place in blood out of the body; but it is to be observed that the degree of aggregation is *always very small* in comparison with what has just been described in the case of the red corpuscles of the blood of man out of the body, or of the bat within the vessels of the living animal.

In the vessels of the web of the frog viewed under the microscope, the red corpuscles, in the natural state, occupy the middle of the stream, and are carried rapidly along, while the colourless corpuscles, like so many small pearls, are seen, especially in the veins and when the flow of blood is languid, accumulated in the lymph spaces, next the wall of the vessel, rolling or sliding sluggishly along, or actually adhering—altogether like what has just been described as seen in the blood vessels of the bat's wing. I never observed the escape of a corpuscle, red or colourless, through a pore of an unruptured vessel in the frog any more than in the bat.†

* For delineations of this, see Plate VI., Fig. 2, and Plate VII. in my essay in Guy's Hospital Reports for 1850, *ut supra.*

† For delineations of all these appearances see Plate IV. in the work cited, and the woodcuts, *passim*, in the text.

PHENOMENA OF INFLAMMATION.

The first step in a traumatic inflammation is *conges-tion;* the second, *increased exudation.* Congestion is manifested by preternatural redness and heat, which are owing to an increased afflux of blood, with accumu-lation of red corpuscles in the small vessels. Exudation is manifested in different ways, according to the struc-ture of the part affected. Swelling, thickening, opacity, phlyctenulæ, pustules, discharge, are appearances in inflamed parts, it is to be remarked, which are owing to the difference in the conditions under which the exuded matter is being or has become converted into cells.

As congestion and exudation are the essential steps of inflammation, so re-establishment of the circulation to its normal condition and the removal of the exuded matter,—that is, the breaking up of the newly formed cellular elements, and the solution and absorption of their remains as indicated by the disappearance of the swelling, &c., constitute the most direct termination in recovery, viz., *resolution* of the inflammation.

In cases in which there is solution of continuity, the healing process is not so direct, and the exuded matter and cellular elements developed therefrom play a more important part. New tissues are formed, by which the wound is united or filled up according as the circum-stances are such as to permit of *"adhesion"* or *"healing by the first intention,"* or only of *"granu-lation* with *suppuration,"* that is, *"healing by the*

K

second intention." The congestion and exudation, and, therefore, the redness and swelling, or the like, of the part do not cease on the commencement of either of these forms of the healing process. They continue as conditions of the process until the wound is united or filled up and cicatrization is completed,—the congestion being the necessary condition of exudation ; the exudation, again, that of the supply of material wherewith regeneration by cell-formation is effected.

ANALYSIS OF THE PHENOMENA OF INFLAMMATION.

Pustular inflammation of the conjunctiva affords an example of inflammation and the healing process in the human body, in their simplest manifestations. It will be useful, therefore, to analyse its phenomena in illustration of those of inflammation in general.

In pustular inflammation of the conjunctiva there is a small, often apparently isolated, spot of vascular injection, and slight swelling of the sclerotic conjunctiva at some little distance from the margin of the cornea, with a flake of matter in the middle. Such are the manifestations of the congestion and consequent exudation. In consequence of the little cohesion of its component cells, the epithelium of the *conjunctiva scleroticæ* does not, like the epidermis of the skin, retain the matter exuded underneath it in the form of a phlyctenula or pustule, but gives way, leaving an abrasion covered with the flake of pus or puriform matter.

Inflammatory redness, as observed in the web of the

bat's wing or frog's foot, is seen under the microscope to be owing to an accumulation or *congestion of red corpuscles* in the blood, within the minute vessels of the affected part. The blood, loaded with the accumulated red corpuscles, may in some of the vessels be still flowing, though sluggishly, like a thick viscid matter ; but in most of the vessels at the focus of the inflammation the red corpuscles are closely agglomerated together, and, being adherent to the walls of the vessels, block them up, so that the blood is stagnant therein.

Here let it be explained in what sense the terms " CONGESTION " and " STAGNATION " are employed. In both cases it is to be understood that there is an *unusual accumulation of red corpuscles in the blood* within the affected vessels. As long as the blood, thus loaded with red corpuscles, continues to flow, however tardily, we have " CONGESTION ;" when the blood has ceased to flow anywhere in the part affected, in consequence of the red corpuscles having agglomerated into masses blocking up the vessels, we have " STAGNATION." *

The vascular injection of the inflamed conjunctiva in man, as observed with the naked eye, or by means of a magnifying glass, resembles in its characters the vascular injection in inflammation of the web of the bat's wing or frog's foot, as observed by the same means. For this reason, and seeing that the plan of distribution

* Essay in Guy's Hospital Reports for 1850, pp. 22-3.

of the smallest vessels in man appears to be similar to that of the distribution of the smallest vessels in the bat or frog; moreover, knowing that the red corpuscles of the blood of man agglomerate together as readily as those of the bat, and much more readily and closely than those of the blood of the frog, we may fairly conclude that in the vessels of the conjunctiva in a case of pustular ophthalmia, the red corpuscles of the blood are accumulated and aggregated together in a manner similar to that in which we can directly observe them accumulated and agglomerated in the vessels of the web of the bat's wing or frog's foot under the microscope.

In hyperæmia or congestion the arteries are dilated, and the blood, loaded with red corpuscles, is still flowing. On such a state of the blood and the blood-vessels, inflammatory stasis or stagnation is liable at any moment to become established. Thus, if constriction of the arteries leading to the affected part take place, and *vis a tergo* thereby be arrested, the red corpuscles with which the last arterial ramifications, the capillaries and venous radicles, are gorged, are observed to aggregate and become stagnant in the vessels. Inflammation may exist independently of an increase in the quantity of fibrine in the blood. Notwithstanding this, it will be easily understood from what has been above stated, that when an increase in the quantity of fibrine in the plasma has taken place, as it tends to do in the course of the disease, the spread and intensity of the inflammation already existing will be thereby promoted, on account of the red corpuscles in such a condition of the blood becoming aggregated both more quickly and closely.

In hyperæmia or congestion of the web of the frog, produced by injury of the spine or section of the ischiatic nerve, I have observed inflammatory stasis in like manner supervene on the occurrence of constriction of the arteries, which I have stated may still be seen occasionally to take place, though in an uncoordinated manner.

On the other hand, immediately after the section of the ischiatic nerve, inflammatory stasis is not so readily excited by the application of salt to the web.* And I have shown in another writing,† that the result of section of the ischiatic nerve was re-establishment of the flow of blood in the web of a frog where there was previously considerable congestion and stagnation of blood corpuscles. This took place in consequence of the resulting dilatation of the arteries permitting the full operation of the action of the heart in propelling the stream of blood in them, against the corpuscles stagnant in their last ramifications, in the capillaries and in the venous radicles, and thus forcing the obstruction. In a similar manner may be explained the effect of section of the nerve leading to the part in the case recorded‡ by

* " Essay on the State of the Blood and the Blood-vessels in Inflammation," *ut supra*, p. 40.

† " Observations on the State of the Blood and the Blood-vessels in Inflammation," in vol.xxxv. of the Medico-Chirurgical Transactions, 1853.

‡ Holscher's " Annalen," Bd. I., p. 498.

Haussmann, of inflammation of the coffin-bone of the horse.

When, however, inflammatory stasis does become established, from any cause, such as a wound, in a frog in which the ischiatic nerve has been divided, it is liable to take place in a greater degree. Thus I have shown, that when the ischiatic nerve has been cut, there is more disposition to congestion below a wound involving division of an artery or vein. And I have shown the cause of this to be the fuller and more rapid flow of blood in the arteries of the web generally, allowing a greater quantity to be poured into the vessels below the wound where the blood stagnates from being in a great measure out of the way of *vis a tergo*. On the same principle, I explained how it is that in an inflamed part of the human body, notwithstanding the congestion and stagnation, an increased quantity of blood enters and circulates. The part, I contended, which is the actual seat of the inflammatory stasis, is to be distinguished from the surrounding parts of the affected organ. It is into and in the latter that the greater quantity of blood enters and circulates, and this because the arteries leading to and distributed in them are relaxed and dilated, without being otherwise affected. To this relaxation and dilatation of the arteries, it is to be added, the more forcible pulsation which characterises them is owing.

The thickening of the conjunctiva in the situation of the injected spot in pustular ophthalmia, and especially the pustule or flake of matter in the middle, are manifestations of the exudation which has supervened on the inflammatory injection.

As to the healing process in pustular inflammation of the conjunctiva:—In cases left to themselves we often find, along with increase of the inflammatory congestion, the abrasion or superficial ulceration of the conjunctiva extended. But under the influence of irritating applications to the eye, the vascular injection speedily disappears, and, *pari passu*, healing of the abrasion or superficial ulceration takes place, the surface becoming covered with a new epithelium, whilst the flake of puriform matter is thrown off.

In regard to the mode of action of irritating applications in conjunctival inflammation, the following explanation is given in my Essay on the state of the blood and the blood-vessels, so often cited :—We have above seen reason to conjecture that inflammation of the conjunctiva, for example, from cold or from the irritation of a foreign particle in the eye, commences by constriction of the small arteries of the affected spot, which allows of the blood corpuscles to regurgitate from the vessels of the neighbouring unaffected parts, and to accumulate in the capillaries and venous radicles. That in such a case, resolution is owing to dilatation of the arteries leading to the affected spot, and coincident acceleration of the flow of blood in them, we have seen equal reason to conjecture. Nay, we have shown, by experiment on the frog, that dilatation of the arteries and the coincident acceleration of the flow of blood, are the first steps to resolution of inflammation ; an experiment, let it be repeated, which is an interesting illustration of the *modus operandi* of stimulating collyria applied to the eye for the cure of catarrhal ophthalmia.

The dispersion of vascular congestion in the web of the frog after section of the ischiatic nerve, and Haussmann's case of cessation of inflammation in the horse's foot after section of the nerve of the limb mentioned above, it will be seen are parallel phenomena.

This view of the matter, it will be observed, is quite at variance with all previous notions on the subject. Mr. John Hunter said that an agent possessing the power to cause contraction of the vessels would probably be the specific in inflammation. The inflammations which Mr. Hunter had here in view were no doubt inflammations of the ordinary kind, in the focus of which the arteries are, I believe, in the state above described, and the vessels he had in view were no doubt arteries, capillaries, and veins indiscriminately. Nevertheless, there are certain forms of inflammation in which Mr. Hunter's conjecture finds a certain degree of application, as I proceed to explain :—

We have in belladonna, as I was the first to show, an agent possessing the power to cause constriction of the arteries ; and in phlyctenular or scrofulous ophthalmia, an inflammation in which belladonna exerts a beneficial influence, not only in relieving the intolerance of light, but also in subduing the inflammatory action. It has, therefore, appeared to me probable that in scrofulous ophthalmia the arteries of the affected part of the eye are in a state of dilatation, and that the belladonna acts beneficially by virtue of its property of exciting contraction of their walls.*

The character of the visible vascular redness of the conjunctiva in scrofulous ophthalmia, its small degree, and the distribution of the injected vessels, lend support to this opinion, and show that there is not congestion and stagnation of red corpuscles in the extremities of the arteries, in the capillaries, and in the venous radicles,—such as, there can be no doubt, occur in pustular or catarrhal ophthalmia, in which irritating applications operate so beneficially; but that, on the contrary, there is an unduly dilated state of the arteries, with a corresponding acceleration of the flow of blood in the part.

Irritating eye-waters, it may be remarked, are prejudicial in the well pronounced form of scrofulous ophthalmia; while in the well pronounced form of catarrhal ophthalmia, belladonna, or atropia, if not exactly prejudicial, is useless. Nay, venous congestion of the healthy conjunctiva is sometimes induced by belladonna or atropia.

EXUDATION AND CELL-MULTIPLICATION.

By a process of osmose or liquid diffusion through the walls of the capillaries, the new materials supplied by the blood for the nutrition of the different tissues pass out, whilst the blood receives in exchange certain

---

* " Ophthalmic Medicine and Surgery," 3rd edition, p. 145, 1865.

of the old decomposed materials of the tissues, which are to be excreted in the respiratory and secretory processes.

On a peculiar osmotic quality of the membrane forming the walls of the capillaries in each different structure, the particular matter required for its nutrition, given out from the blood, seems to depend ;—differences in the quality of the intervening membrane (the separated fluids remaining unchanged) determining different osmotic effects. A morbid alteration in the osmotic quality of the capillary walls will thus lead to a difference in the materials which shall be diffused out from the blood in the capillaries on the one hand, and which shall be diffused from the juices outside into the blood within the capillaries on the other.

Again, with differences in the quality of the two fluids separated (the intervening membrane remaining the same), different osmotic effects take place. A morbid alteration in the condition of the blood within, or of the juices outside the capillaries, will, therefore, lead to a difference in the nature of the material reciprocally diffused.

Thus it is, no doubt, that the differences in the quantity and quality of the matter exuded from the blood in inflammation, which may exist, are determined.

The matters which pass out from the blood through the pores of the capillary walls, must be in a state of solution at the moment of exudation, how soon soever it may become consolidated and appropriated in the process of cell formation which may characterise the particular form of the inflammation. In like manner, the matters which enter into the blood by diffusion through the pores of the capillary walls in exchange for those passing out from it, must likewise be in a liquid state.

Professor Virchow of Berlin, in his lectures on " Cellular Pathology," insists so much on the process of cell multiplication by division of pre-existing cells, or as he calls it, " proliferation," being the essential part of the inflammatory process, that he appears almost to ignore the connection of inflammation with any disturbance of the circulation at all ; and even to ignore the necessity of any supply of material from the blood for the maintenance of " proliferation," or cell multiplication.

By calling inflammation " Abnormal Nutrition," and so far ignoring vascular congestion as a part of the process, the followers of Virchow seem to think that he has struck out a new view in pathology.

On Professor Virchow's views that hyperæmia, or congestion, exercises no direct influence on the nutrition of the tissues, and that " proliferation " is the representative of the inflammatory process, I would remark :—In normal nutrition, cell formation goes on *pari passu* with the supply of the necessary material from the blood. This supply of nutritive material depends on the circulation of the blood. When that is disturbed, the supply of material is disturbed also, whether by excess or defect ; and the normal process of cell formation is thereby disturbed. Professor Virchow speaks of "proliferation" as if the cells were the *agents* of the inflammatory process,—that being stimulated to action they set to work to *proliferate* more actively than usual, as if they could vary their activity independently of the conditions around them in respect to supply, by the circulation of material in requisite quantity, and of requisite quality.

It would be out of place here to enter into the question as to whether development of cells takes place free in the blastema, or in the interior of pre-existing cells,—the contents of the latter serving as a blastema, in and out of which the endogenous formation goes on,—or whether, as Virchow maintains, the multiplication of cells goes on by "proliferation" or division of pre-existing cells. It is sufficient to admit that the process of cell-formation does go on—no matter how. But in admitting this, we do not admit that cells are the essential agents of the inflammatory process. Cell-formation is only one of a chain of processes. It is dependent on the supply of material, and the supply of material is dependent on the afflux of blood. The cells multiply merely because the materials and conditions for their multiplication exist. There is no self-sufficiency of cells in the sense in which Virchow seems to view the subject.

But admitting for a moment Virchow's view, that the inflammatory process is independent of any disturbance of the circulation, and that it consists essentially in cell-multiplication, the word "proliferation," as involving an hypothesis, is objectionable. I apprehend that in an inflamed part of the living bat or frog, Virchow never exactly saw under the microscope how the cell-multiplication was going on,—whether by a process of free cell-development, by a process of endogenous formation, or by a process of proliferation.

The question raised by Virchow as regards the pathology of inflammation appears thus to reduce itself to—whether cell-multiplication on the one hand, or congestion on the other, be the first step in the inflammatory process. Now, in answer to this question, I can say that, having inflicted a wound in the web of a frog, and watching the development of the inflammation thereby excited, we can see under the microscope that congestion is the first step ; and that in man, so far as our means of observation extend, redness, heat, and swelling—indications of congestion—are the phenomena that a cut finger first presents in the situation of the wound.

Professor Virchow surely cannot have devoted much continuous attention to the observation of the microscopical phenomena of the circulation in reference to inflammation, otherwise he would not have misinterpreted the state of the blood and the blood-vessels in the process in so many particulars, as he has done.

It has been justly remarked that such a view of " Inflammation " is taking away from the word its primitive meaning, and by a loose generalization, imposing on us a pathology, according to which every diseased organ would be regarded as inflamed—every cadaveric lesion, as the result of an inflammation, acute or chronic.*

But Virchow appeals especially to the phenomena of *Keratitis,* or inflammation of the cornea, in support of his view that "proliferation," considered as the essential part of the inflammatory process, is independent of hyperæmia, or congestion ; the cornea being a non-vascular structure. Let us, then, examine the phenomena of Keratitis, or inflammation of the cornea.

KERATITIS, OR INFLAMMATION OF THE CORNEA.

Structure of the proper substance of the cornea.—A clear fibrillated, but otherwise homogeneous-looking, intercellular substance, in the form of laminar fibres, constitutes the groundwork of the cornea. In the interstices between the fibres are the elongated remains of cells with nucleus fibres. These interstices anastomose with each other, and form a network throughout the corneal structure. By maceration of the cornea, its substance becomes swollen and spongy-looking, from imbibition of water into the interstices. This shows that the laminar fibres of intercellular substance are arranged together somewhat in the manner of the fibrous framework of a compressed sponge, and are not aggregated so as to form separate and distinct layers. It is through its proper substance that the cornea is joined to the sclerotica, the fibres of the two structures interlacing or being continued into each other.

* Marey, *op. cit.* p. 401.

Except in the early stage of development, no blood-vessels are visible in the cornea beyond its margin. But though vessels do not actually ramify in the cornea, this structure, I would observe, is not less dependent on the circulation than other structures of the body. The nutritive material with which its interstices are filled is derived by transudation from the blood circulating in the vessels of the adjacent conjunctiva and sclerotica. · The circumcorneal vascularity of these membranes is evidently greater than their own nourishment requires. It is now a quarter of a century since I insisted on the fact that the inflammatory process in the cornea is not essentially different from that in other structures. I showed that if the old saying, "*Ubi stimulus, ibi fluxus,*" is not applicable to inflammation of the cornea excited by a wound, a modification of it is applicable, viz., "*Hic stimulus, ibi fluxus.*"*

The cornea then, there is reason to believe, derives the materials necessary for its nutrition from the blood circulating in the vessels of the adjoining parts of the conjunctiva and sclerotica. Let us inquire into what takes place in the cornea when it suffers such an injury as would excite inflammation in one of the vascular parts of the eye.

When the cornea is injured, congestion in the vessels of the adjoining parts of the conjunctiva and sclerotica takes place, and exudation into the substance of the cornea at the injured place by-and-by ensues. Thus,

* "British and Foreign Medical Review," No. XXXIV., 1844.

though non-vascular, and of course not the seat of the inflammatory congestion, the cornea becomes the seat of a very important part of the inflammatory process —the most important part, perhaps, as regards the events of the process. The cornea in this state may, therefore, be said to be, to all intents and purposes, inflamed, the only difference in respect to it, as compared with vascular parts, being that the vascular congestion is not *in it*, but in *adjoining structures.*

In the progress of inflammation in the cornea, this structure may become vascular, but such an event is owing to the development of new vessels, such as also happens in inflammation of vascular parts.

Though inflammation of the cornea, considered as a non-vascular part, has been thus dwelt on, the truth is that all tissues, as regards their component elements, are, properly speaking, non-vascular, and differ from the cornea only in the degree of proximity to the vessels, and in inflammation only in the degree of proximity to the source of the exudation. But this very difference in the case of the cornea affords a natural analysis of the inflammatory process. It enables us to observe, separately, the two great stages of inflammation proper —the *congestion* and *exudation*—the congestion in one place, the exudation in another. It also enables us to observe, in an uncomplicated manner, the eventual stages of inflammation, such as reorganization and suppuration.

That the cornea is the seat of exudation is manifested by opacity of various kinds, phlyctenulæ and abscesses. When new vessels are developed in and out of the exuded matter, the cornea then becomes the seat of more or less redness.

The cornea has been just referred to as a part which, from its non-vascularity, forms a good subject for observing, in a manner uncomplicated by the presence of vessels, the manifestations of the conversion of the exuded matter into cells in the healing process. The matter exuded into the interstices of the cornea from the vessels of the adjacent conjunctiva and sclerotica may (as far as examination with the naked eye, or the eye assisted merely by a magnifying glass, goes), be seen to undergo the different modes of development above described. It may be seen that pustules form, and that, in the healing of incisions and ulcers of the cornea, adhesion and granulation, and, lastly, cicatrization, may take place without the development of new vessels—a convincing proof that neither suppuration nor "organization" necessarily implies the development of new vessels.

But, as above hinted, new vessels may be formed in the matter exuded into the cornea, and this again affords a very interesting and readily observable example of the development of new vessels uncomplicated by the previous existence of other vessels in the part.

The mode in which congestion of the vessels of the adjacent conjunctiva and sclerotica is excited by injury of the cornea alone seems to be by irritation of the nerves of the part (for the cornea has nerves, though no vessels) making, by reflex action, an impression on

L

the vaso-motor nerves of the arteries of the adjacent parts mentioned—*hic stimulus ibi fluxus*—just as a foreign particle in the eye excites inflammatory congestion of the conjunctiva.

The idiopathic inflammation of the cornea, known by the name of *keratitis parenchymatosa*, has been viewed as primary, on account of the slight appearance of increased vascularity at first seen in the adjacent white of the eye. But the increase of vascularity, small as it is, is quite in proportion to the degree of morbid alteration in the cornea. If at the very outset of the inflammation the initial step be a morbid change in the cornea, this is immediately followed by an afflux of blood in the white of the eye, like what we have seen supervenes on injury of the cornea.

Having shown that inflammation and the healing process in the cornea are dependent on essentially the same conditions as those on which they depend in parts more especially reputed vascular, I will now quote, in conclusion of this chapter on inflammation, from the essay I have ventured so often to refer to, the account of my observations on the healing process of wounds in the frog's web.

OF THE STATE OF THE BLOOD AND THE BLOOD-VESSELS DURING THE HEALING OF A WOUND OF THE WEB OF THE FROG.

When an artery only is cut across.—In a day or two after the injury, sooner or later according to circum-

stances, the flow of blood usually becomes re-established
in the vessels in which there may have been stagnation,
but it goes on more or less sluggishly. In two cases
the cut ends of the artery became reunited, and the
flow of blood was restored as before.

When the wound of the web is small, it becomes
filled up with a grayish substance, composed of cells
with granular contents. In this substance, by which
the wound may be closed in two or three days, no circulation is at first seen ; but in the course of a fortnight or three weeks red blood corpuscles may sometimes
be observed obscurely threading their way in a channel
or channels in the opaque, newly-formed substance.

When the wound of the web is large, immediate
union does not take place. The edges of the wound
become thickened, and the capillaries all round, but
especially those on the distal side, much congested
and enlarged. The newly formed substance, which is
composed of cells in process of being metamorphosed
into cellular tissue, appears at the edge of the wound
in the form of a clear border of fibrous-looking tissue,
without blood-vessels. In this substance, however,
capillaries appear to be by-and-by developed in connection with the surrounding old ones ; but before this
is accomplished more new substance has been added at
the edges, so that they still continue to present a clear
non-vascular border. The capillary network around
the wound receives its blood from the branches of the
artery above, and, in a retrograde direction, from the
artery below. This process goes on until the granu-

lating edges meet, when the non-vascular substance coalesces, and the wound is closed. The appearance of the part now is that of a gray non-vascular spot in the middle, with a vascular network around.

When capillaries only are cut across.—The capillaries actually cut shrink and disappear, whilst those with which they communicated become somewhat enlarged, and form a marginal network around the wound.

When a vein only is cut across.—As in the case of section of an artery, the wound of the web, when small, may become closed in the course of two or three days, by a grayish granular substance interspersed with pigment granules. By-and-by, in channels in this new substance, red blood corpuscles may be observed threading their way, though obscurely on account of the opacity of the new substance.

When a larger wound has been made in the web, immediate union does not take place ; but the process analogous to granulation, above described in the case of section of an artery, is established at its edges.

When arteries, capillaries, and veins are cut across. —In this case the wound is necessarily of considerable size. Immediate union, therefore, does not take place.

Around such wounds inflicted about a week before, inflammatory redness was visible to the naked eye. Under the microscope it was observed that the capillaries in the red part around the wounds were gorged with red corpuscles, and that the blood was for the most part flowing, though more or less sluggishly. No new vessels were as yet formed. The blood, however, flowed in a retrograde direction in certain of the old vessels. The particular connection of the capillaries around the wound with the nearest branches of the cut arteries and veins, both above and below the wounds, will be understood from what has been already said under the three preceding heads.

I have sometimes found the lower part of the cut artery tortuous and dilated at intervals.

At the extreme margin of the wound the process of granulation, including the development of new capillaries, goes on in the manner above described, as when an artery only is cut across. The new capillaries constitute the terminal loop of the network margining the wound. Here the blood is stagnant, but in the rest of the network the blood, loaded with red corpuscles, is seen flowing, though more or less sluggishly.

The edges of the wound where the granulating process is going on are thick from exudation in the substance of the web, and become drawn together so much, that if the wound is not too large, they at last meet and unite. After this many of the capillaries shrink and disappear, and the circulation becomes freer in the remainder. The cicatrice continues thick for a while, and the web is puckered around it. If the wound be large it does not close, but its edges cicatrize.

CHAPTER XIII.

IN certain morbid states of the heart a deposition of coagulated fibrin takes place from and in the blood in the interior of that organ. In phlebitis, a thrombus, or clot, is found filling up the cavity of the vein at the part affected. In inflammation of an artery also a clot may be found. Abrasion of the lining membrane of the heart or vessels from inflammation is a condition for such deposition of coagulated fibrin at the place ; and a morbid state of the blood favours the occurrence.

That traumatic abrasion of the lining membrane of a vessel in a healthy animal is alone a sufficient condition, however, for the deposition of fibrin by coagulation from the blood at the place is proved by the following experiments, the results of which on the frog are described and delineated in my essay on inflammation in Guy's Hospital Reports for 1850, pp. 42 and 58, and the results of which on the bat are described and delineated in my paper on the veins of the bat's wing in the Philosophical Transactions for

1852, pp. 133-4. In the frog it was found that, on pressing the web over an artery or vein—a large vein especially—pretty firmly with a blunt point, a flocculent deposit occurred on the inner surface of the wall of the vessel at the place injured, more or less completely obstructing its channel. The floccules, which were composed of coagulated fibrin, with blood corpuscles entangled in its meshes, chiefly colourless corpuscles, but also a few red ones, were after a time gradually detached in fragments, and carried away in the stream of blood. In the bat, when pressure was applied in a similar manner over an artery and vein of the web of the wing, a deposit of a viscid-looking grayish granular lymph within the vessel at the place occurred, obstructing its channel, and narrowing the stream of blood (*loc. cit.*, figs. 3 and 4). On watching, I have seen portions of this deposit become detached, and carried away by the stream of blood, with corresponding enlargement of the channel, and again a new deposit take place, with renewed narrowing of the stream. When the pressure has been considerable, I have seen the vein become for a time wholly obstructed by the deposit. A similar deposit of lymph occurred in the artery.

Flakes becoming detached from fibrinous deposits in the heart, or from phlebitic clots, and carried along in the stream of blood, are at last arrested in their course in arterial branches, the caliber of which is too narrow to give them passage. The flake of fibrin thus forms an embolon, or plug, which blocks up the channel of the artery. This blocking up of the artery has been named *embolism*. In phlebitis the flakes of fibrin, or puro-lymph, detached from the thrombus or clot in the affected veins, becoming arrested in small arteries of the lungs or liver, there give rise to circumscribed inflammations and abscesses, which are correctly viewed as *metastatic*.

But the question now arises what is the mode in which the circulation of the blood is disturbed by an embolon in an artery ?

The effects of embolism are commonly described as a collapse and emptying of the artery below the embolon, and of the capillaries and veins to which the artery leads. That this is a misconception as regards the arteries of the web of the frog, at least, appears from my microscopical observations on the subject. In the web of a frog under the microscope, we some-times see an artery become blocked up by a mass composed, apparently, of colourless corpuscles and coag-ulated fibrin. In a web, in the vessels of which stagna-tion of blood had been produced by the action of salt, an artery, in which the flow had continued free, sud-denly became stopped up by a plug of gray granulous substance, so that the blood was there arrested in its course, the stream passing off by the first considerable branch above the obstruction. Below the obstruction, the blood flowed into the artery in a retrograde direc-tion by one set of branches, and passed out from it in a direct course by another set of branches, as in the case of interruption of the flow of blood in an artery by division of the vessel. By-and-by the plug was pushed along in the artery by the force from behind, and the flow of blood was re-established down to the first consider-able branch above the place, where the plug again stopped. A portion of the plug of granulous substance becoming detached and carried away, the mass was reduced in size, so that it admitted of being again forced on. It next caused obstruction in a capillary; but while it was in the act of being broken up and its fragments carried away in the stream, the struggles of the animal interrupted the observation. In another case, one in which a large portion of the web had sloughed away in consequence of inflammation, excited two or three days before, I several times saw, while watch-ing the flow of blood in the remaining part of the web, capillaries suddenly become filled with a gray granulous looking substance. This, for a time, obstructed the vessel; but it was always at last moved on by *vis a tergo*, and getting into a vein was carried away. In other cases I have seen similar masses adhering to the wall of the vessel, but not entirely stopping it up. In some cases I have been able to satisfy myself, that the plugs of gray granular substance consisted of colour-less corpuscles, agglomerated and held together by a tenacious looking matter, probably coagulated fibrin. In other cases, it appeared to me that the gray sub-stance consisted of minute granules, held together by the tenacious matter.

The following observation, which I record for the first time, belongs to the same category as those just

quoted from my essay in Guy's Hospital Reports :—
A frog, which had been subjected to section of the
ischiatic nerve on one side some weeks before, and
consequent on which destruction of the webs of the
corresponding foot had taken place by ulceration, was
the subject of the observation, and the web of the
uninjured leg the part examined. The arteries were
seen to be in a medium state of width. At intervals
of a moment or two a mass of agglomerated red
corpuscles was seen to rush along in a flood. After
the mass of red corpuscles had passed, the artery
appeared at first sight as if empty at the place, but
on closer examination it was seen that there were
here plasma and colourless corpuscles.* These agglo-
merations of red corpuscles in the course of the circulation had come, I believe, from the vessels in the
ulcerated foot of the opposite side. For there the
blood flowed slowly like a viscid red mass, so closely
aggregated were the red corpuscles.

In like manner, the flakes of gray granulous substance which were seen to block up the vessels in the
other observations above described, were derived no
doubt from deposits in the vessels of the inflamed and
ulcerated parts.

The disturbance of the circulation produced by the embolon in the artery in one of the cases above mentioned is

---

* The phenomenon reminded me of the appearance of drops of
red blood passing along in the caudal vein of the eel, as if propelled
from the caudal heart, as described in the paper in the Philosophical
Transactions for 1868, above referred to at p. 100.

likened to that which is caused by the section of an artery. The effect of section of an artery of the web of the frog on the flow of blood in the part is thus described at p. 23 of my essay in Guy's Hospital Reports for 1850 :—" When an artery is cut across, it immediately becomes constricted, even to closure of its caliber, upwards in the direction of its trunk, and downwards in the direction of its ultimate ramifications. The flow of blood is thus arrested, and the immediate consequence is, an exsanguine state of the part to which the ramifications of the artery lead. But this state of matters is not of long duration. In the course of a minute or so, relaxation of the wall of the artery and dilatation of its caliber take place both above and below the wound. In the upper part of the artery, the flow of blood is re-established as far down as the first considerable branch proceeding from it above the place of section. By this branch the stream of blood passes off. In general, none of the blood, except a stray corpuscle now and then, enters the artery further, although it has become dilated down to the place of section, where, however, the cut end of the vessel continues closed by constriction. Into the part of the artery below the section blood, of course, no longer enters directly. It enters, however, in a retrograde direction, and very slowly, by one set of branches, and passes out in a direct course, but still very slowly, by another set of branches. The blood, which enters the artery below the section in a retrograde direction, regurgitates from the capillaries and veins to which the branches lead by which the blood enters ; and, if the cut artery has a direct anastomoses below the section with another artery, blood also regurgitates by that anastomoses. The branches by which the blood enters in a retrograde direction, and those by which it passes out in a direct course, vary according to circumstances." For examples and figures illustrative of this, see Guy's Hospital Reports for October, 1850, pp. 24, 25, 26, as regards the arteries of the frog's web.

According to the part of its course where an artery is blocked up by an embolon, so will be the number and extent of parts affected by the stoppage of the flow of blood to them ; and according to the number and freedom of anastomoses with adjacent arteries above and below the place where the embolon is lodged in the affected artery, so will be the supply of blood to the parts below and the freedom with which it will circulate.

In illustration of all this, experience of the cases in which an artery has been tied is applicable.

The central artery of the retina not having any free anastomoses with other arteries, embolism of it must be a cause of serious interruption of the circulation in the retina, but I question much whether the cases described as cases of embolism of the central artery of the retina were really of that nature, or at least wholly so. The following is an outline of the symptoms and appearances in such cases :—

With sudden failure of sight, the optic disc has been found pale, the arteries constricted, and the veins dark

and rather large-looking, except at their exit in the
disc, and at one or two other parts of their course
where they were small and collapsed. The constricted
arteries were not observed to pulsate, even on making
strong pressure on the eyeball. In the region of the
macula lutea the retina was somewhat swollen, and
presented a milk-white opacity pervaded by small
isolated-looking vessels or streaks of ecchymosis. The
macula itself has sometimes been obscured in the
opacity, but occasionally seen very distinct and red in
the middle.

After a time the arteries have been found dilated
somewhat and again exhibiting pulsations. The swelling and opacity of the retina in the region of the
macula, also, have diminished or disappeared. The
event, however, has generally been well marked white
atrophy of the disc with confirmed amaurosis.

In this description there is *nothing pointing to an
embolon in the central artery*. The state of the
vessels is merely such as would arise from spasmodic
constriction of the artery; and such constriction is just
what might occur suddenly and determine the failure
of sight by stoppage of the afflux of arterial blood to
the already morbid retina. For it is to be observed
that the whiteness of the disc and swelling and opacity
of the retina in the region of the macula lutea could
not have been of the same recent occurrence as the
sudden failure of sight.

The probability seems to be that deterioration of the
structure of the optic nerve and retina had been

gradually taking place in the one eye, with corresponding impairment of sight, though as yet undetected by
the patient, when the spasmodic constriction of the
central artery, supervening suddenly, aggravated the
blindness and drew the patient's attention to it.

Heart disease has been generally, though not in-
variably, detected in such cases ; but embolism of the
vessels of the intracranial part of the optic nervous
apparatus would be quite cause enough of the disturbance of the circulation in the eye, without appealing to
an embolon in the central artery of the retina itself.
If, in any of the cases, there was really a plugging of
that artery, the peccant matter was probably softened
puriform fibrin, such as Mr. Gulliver* has described in
clots of the heart, arteries, and veins, and originally
discovered, by an extensive series of experiments and
observations, to be a distinct and important elementary
form of disease. Long after the publication of Mr.
Gulliver's observations, the results were described anew
under the name of " thrombosis."†

* " Medico-Chirurgical Transactions," 1839, vol. 22 ; " Edin. Med.
and Surg. Journal," vol. 60 ; Dr. John Davy, on " Diseases of the
Army," p. 267.

† In " The Nomenclature of Diseases drawn up by a Joint Com-
mittee appointed by the Royal College of Physicians of London,"
1869, we find, at pp. 35 and 63, as follows :—1. Thrombosis (local
coagulation) ; 2. Embolism (coagula conveyed from a distance).

CHAPTER XIV.

INQUIRY INTO THE NATURE OF THE DISTURBANCE OF
THE CIRCULATION OF THE BLOOD IN THE OPTIC
NERVOUS APPARATUS, OCCASIONED BY IMPLICA-
TION OF THE ROOTS OF THE SYMPATHETIC IN
DISEASE OR INJURY OF THE CERVICO-DORSAL
PORTION OF THE SPINAL CORD.

THE points treated of in the five preceding chapters
have so significant a bearing on our subject, that I
have not considered it too much to dwell on their
elucidation at the length I have done, the more especially as in the case of certain of them, great misconception, it has been shown, prevails. It is surprising,
indeed, to see that while the mechanism of the circulation through the heart and great vessels is so well
known, the mechanism of the circulation in the extreme vessels is really imperfectly understood, notwithstanding that the subject is one of fundamental
importance in pathology as well as in physiology. In
the pathology of inflammation, especially, we have seen
what an important link in the chain of processes the
state of the blood and blood-vessels forms.

To resume, now, our special investigation.

The sympathetic nerve governs the contractions of the walls of the arteries and so regulates the variations in the width of these vessels. Variations in the width of the arteries of an organ imply, it is to be remem-· bered, modifications in the flow of blood in the part, independently of the general effect of the heart's action. Thus it is that the healthy circulation in the optic nervous apparatus, in the eyes and in certain other parts of the head, depends on the integrity of the sympathetic nerves in the neck and of their roots in the spinal cord, whereas disturbance of the circulation in those parts results from lesion of the sympathetic nerves and their roots in the places mentioned. The effect of such a disturbance of the circulation is to impair the vital energy of the parts implicated, and to pervert the nutritive processes therein in such a manner as to lead, eventually, to degeneration of structure and consequent interruption of function.

In the cases related in Chapter II., the roots of the sympathetic in the neck must necessarily have participated in the morbid state of the spinal cord induced by the concussion, in the accidents on which the failure of sight supervened. The effect has been the disturbance of the circulation in the optic nervous apparatus, from which the impairment of sight has directly resulted.

Irritation of the roots of the sympathetic nerve in the cervico-dorsal region of the spinal cord from morbid action going on there, excites contraction of the circular muscular fibres of the arteries of the optic nervous apparatus, &c., which derive their vaso-motor nerves therefrom, and consequent constriction of the caliber of those vessels. The effect of this is impeded circulation in the parts, as above explained at pages 77—79.

Actual injury of the roots of the sympathetic nerve in the cervico-dorsal region of the spinal cord, on the contrary, has for effect paralysis of the circular muscular fibres of the arteries, and consequent dilatation of the caliber of those vessels. The effect of this is the same as that which, we have seen, results from section of the sympathetic nerve in the neck, namely, a determination of blood to the parts, in consequence of a fuller and freer flow in the arteries (page 102).

Similar effects were exemplified in cases of locomotor ataxy, reported by Professor Trousseau,[*] Dr. Duchenne,[†] and Dr. Bland Radclyffe.[‡] Thus, when the patient was free from pain, the conjunctiva was bloodshot ; but during a paroxysm, when the pain reached a certain degree of severity, the conjunctiva was observed to become free from the vascular injection of which it was the seat during the intervals between the paroxysms. Along with the cessation of the vascular injection the pupil became dilated.

The explanation to be offered of these phenomena is, that during the paroxysm of pain the roots of the

---

[*] *Ut supra.*

[†] In Trousseau, *ut supra.*

[‡] In Dr. Russell Reynolds' "System of Medicine," vol ii. p. 346.

sympathetic in the cervico-dorsal region of the spinal cord were, in common with the roots of the sensitive nerves, in a state of irritation, and the consequence was, 1st, excitement to contraction of the circular muscular fibres of the small arteries of the conjunctiva and other parts of the eye, producing constriction of those vessels and impeded flow of blood; and, 2nd, excitement to contraction of the radiating muscular fibres of the iris, the result of which was the dilatation of the pupil. In the intervals between the paroxysms, the energy of the vaso-motor nerves being exhausted the muscular fibres under their governance became relaxed. Hence the dilatation of the arteries with afflux of blood and the contraction of the pupil from the unrestrained action of the sphincter.

Dr. Brown-Sequard * has lately dwelt on the frequent occurrence of vascular injection, with an elevated temperature, and increased sensibility of the conjunctiva, face, and neck, of a half-closed state of the eyelids, and of a contracted state of the pupil, in cases of spinal hemiplegia from injury of the spine in the cervico-dorsal region on the side corresponding to the lesion.

In our cases the roots of the sympathetic nerve in the neck, from which the arteries of the optic nervous apparatus derive their vaso-motor nerves, were not affected in the same degree. They were weakened and irritated rather than paralysed, somewhat as in loco-

---

* Lectures in the *Lancet* of November the 7th and 21st and December the 12th, 1868.

motor ataxy, and the result on the circulation was the disturbance arising from unregulated action of the arteries leading to a low form of inflammatory congestion, as will be explained in the next chapter.

CHAPTER XV.

BLOOD-VESSELS OF THE OPTIC NERVOUS APPARATUS.

PREPARATORY to tracing the progress of inflammation from within the cranium, along the optic nerve to
the optic disc, it will be useful to recall to remembrance
the distribution of the blood-vessels of the optic nervous
apparatus.

The intracranial and intraorbital parts of the optic
nervous apparatus receive their arterial ramifications
from arteries adjacent to them in the different parts of
their course. These arterial ramifications are accompanied by corresponding veins. The capillary network
with the extreme arterial ramifications opening into it,
and the venous radicles arising from it, forms a continuous vascular system from the central to the peripheral end of the optic nerves—from the tubercula
quadrigemina to the optic discs.

The vessels of the retina form a system distinct
from that of the other parts of the optic nervous
apparatus.

The fibres of the optic nerve visible to the naked eye
are fasciculi of microscopical primitive fibrils, enclosed
each in a neurilemma or cellular sheath, and the whole
are enveloped in a common neurilemma. The neu-
rilemmata of the fibres cease as the nerve penetrates
the sclerotica, whence arises the constriction of the
nerve at that place, and the structural arrangement
commonly described under the name of *cribriform
lamina* of the sclerotica. There, free from neurilemma,
the fibres of the optic nerve pass through a well-
defined opening in the choroid to join the retina.
About the third of an inch from the sclerotica the
optic nerve is perforated obliquely from below by the
central artery and vein of the retina, which run in a
canal in the axis of the nerve, to gain the interior of
the eye, where they ramify on the inside of the retina.

The small vessels which, when filled with blood
loaded with red corpuscles, give redness to the optic
disc, belong both to the vascular system proper to the
optic nerve, and to the vascular system proper to the
retina. These small vessels of the optic nerve and
retinal systems anastomose with each other on the disc,
and hence when they are much congested at the same
time, the redness of the disc and the redness of the
retina run into each other, so that the boundary of the
disc is lost in indistinctness.

The central artery of the retina is a branch of the
ophthalmic, which also gives off the ciliary arteries,
and the arteries of the various accessory parts of the
eye. The ophthalmic artery itself arises from the in-

ternal carotid on its emergence from the cavernous sinus.

The central vein of the retina, the ciliary and vortiose veins, the veins of the ocular conjunctiva, &c., end in forming the ophthalmic vein which enters the cranium by the sphenoidal fissure, and opens into the forepart of the cavernous sinus. The ophthalmic vein maintains a free communication between the veins of the scalp and face and the cranial sinuses.

The sinuses of the dura mater are venous canals, which, from their structure, are incapable of diminution in their width by contraction of their walls or any material dilatation by distention with blood from within. The two cavernous sinuses communicate with each other through the circular sinuses,—one of which lies in front and the other behind the chiasma of the optic nerves. Both the circular and cavernous sinuses receive veins from the adjacent parts of the brain and meninges. From the cavernous sinuses the blood is carried by the petrosal to the lateral sinuses, which open into the internal jugular vein. All the sinuses communicate freely with each other.

INFLAMMATION OF THE OPTIC NERVE AND DISC.

In intracranial disease (basilar meningitis, tumours, hydrocephalus) inflammation may be propagated along the optic nerve, and come to manifest itself in the eye, under the ophthalmoscope, in the form of inflammatory injection and slight swelling of the optic disc, with some implication of the retina. Or, instead of inflammation being propagated along the optic nerve, through its system of vessels, the intracranial disease may come to manifest itself through the retinal system of vessels within the eye under the form of venous congestion and prominent swelling of the optic disc. Or, the eye may become implicated in both ways.

It may happen that inflammation, commencing in the intracranial part of the optic nerve, does not extend down so far as the optic disc, or, at least, not until some considerable time ; but it will operate, nevertheless, as a break to the transmission of impressions from the retina to the brain. Hence, there may be failure of sight, with little or no appearance of anything abnormal under the ophthalmoscope.*

Inflammation of the optic disc presents itself in various degrees from what is called simple hyperæmia, or congestion, to what is called acute neuritis. The vascular injection at first implicates the nasal side of the disc and adjacent part of the retina most. The congestion spreading, the whole disc is found red, so that the distinction between it and the retina is lost, and its locality is ascertainable only by tracing the

---

* See in further confirmation of this :—" Kürzere Abhandlungen, Notizen und casuistische Mittheilungen vermischten Inhalts," von Professor A. von Graefe. S. 2 ; "Ueber Neuro-retinitis und gewisse Fälle fulminirender Erblindung," p. 128 ; " In Archiv für Ophthalmologie, Zwölften Band,—Abtheilung, II., Berlin, 1866." Professor von Graefe here well remarks, that in *neuritis fulminans*, the papilla may not be implicated until some time after the trunk of the optic nerve has become affected. Hence we may find great functional disturbance with little appearance of anything abnormal in the papilla.

retinal vessels to the point where they emerge. This part may be as yet white, and free from redness. Along with the congestion there is some degree of swelling of the disc. In a more advanced stage there is mixed with the redness of congestion grey opacity extending to a greater or less extent on the retina from exudation. Sometimes spots of extravasation present themselves.

In *venous congestion* of the optic disc there is, with intense redness and dark grey opacity, swelling so great that the disc is prominent above the surface of the fundus—abruptly prominent on one side, and shelving down on the other. It may thus admit of being seen under the ophthalmoscope by the erect image exploration. The retinal arteries are constricted, but the veins very turgid and tortuous; here and there patches of extravasation may be observed. The contour of the congested disc is indistinct, and the adjacent part of the retina somewhat opaque.

Notwithstanding this affection of both disc and retina, the patient may still retain a considerable amount of vision for a time, from which it may be inferred that the extra-ocular part of the optic-nerve is not at first involved in the disease.

It has been conjectured that venous congestion of the intra-ocular end of the optic nerve is owing to pressure on the central vein of the retina causing an impediment to the free return of blood from the eye by that vessel. This, Professor A. von Graefe * attributes to an

* In " Archiv für Ophthalmologie," *ut supra.*

" incarcerating action " of the unyielding sclerotic ring
on the central vessels of the retina, as they run enclosed
within the optic nerve, at its entrance into the eyeball.
An additional cause in operation, though admitted to
be only auxiliary, is supposed by Dr. T. Clifford Allbutt*
to be pressure on the cavernous sinus by meningitic
exudation at the base of the brain, or other implication
of the cavernous sinus in the inflammation, whereby
"the ebb of the blood from the ophthalmic vein is
slackened."

To account for the congestion and swelling of the
intra-ocular end of the optic nerve, it is scarcely necessary to invoke any such cause as pressure on the vessels.
Simple constriction of the central artery of the retina,
from contraction of its muscular walls—the co-existence
of which with turgidity of the veins von Graefe so well
recognises—would be, it is probable from what is said
at page 132, a sufficient cause of the accumulation and
stagnation of blood in the venous radicles with the
consequent exudation into the intra-ocular end of the
optic nerve, causing the great swelling. Thus, if while
the flow of blood in a part is full and free, the artery or
arteries leading to it become constricted, aggregation
and stagnation of the red corpuscles take place in the
capillaries and venous radicles. Constriction of the
arteries now becoming less, blood is gradually allowed to
pass, but in consequence of the weakness of *vis a tergo,*

* On the Diagnostic value of the Ophthalmoscope in "Tubercular" Meningitis, in the *Medical Times and Gazette,* for May 1,
1869, p. 597.

it is not driven on, but accumulates in and distends the veins in the manner we see them. In University College Hospital, under Sir William Jenner, there was lately a boy, six years of age, affected, as was found after death, with a tumour of the Pons, in whom, besides venous congestion of the disc and retina, there was great venous congestion of the conjunctiva of the right eye, with ulceration of the cornea and puro-mucous secretion. Here the venous congestion of the conjunctiva at least could not be attributed to any incarcerating action of the sclerotic ring.

It may be suggested that the sclerotic ring with the cribriform lamina is an anatomical arrangement calcu-lated rather to guard against incarceration of the retinal vessels than to promote it. If such incarceration were liable to occur, a similar impediment to the escape of blood from the choroid by the trunks of the vorticose veins might be expected. But here we can see that the mechanism is rather to guard against such an accident.

In the cases of spinal concussion which we have under consideration, the effect as regards the eyes, has been a disturbed state of the circulation in the optic nervous apparatus, passing into a low form of inflammatory action, which is leading, or has already led to perverted nutrition and degenera-tion of structure, with a corresponding degree of amaurosis or amaurotic amblyopia, and its attendant symptoms.

The objective appearances, so far as the optic disc and retina are concerned, I have already described as indicative of hyperæmia, or congestion of those parts. What might be the appearances of the state of the blood and the blood-vessels in the fundus of the eye under a higher microscopic power than that under which we can view the optic disc and retina by means of the ophthalmoscope, of course cannot be known, but I venture to offer, by way of illustration, the microscopical observations above related at page 107, on the state of the blood and the blood-vessels in the web of the foot of a frog which had suffered an injury of the spine. In this state, the distended dark-red veins were very evident to the naked eye, but not so the gorged capillaries. With a magnifying-glass a red glimmer only could be detected in the web where they were. Such enlargement of the veins from engorgement with red blood corpuscles we may thus, it will be seen, accept as a naked-eye indication of capillary congestion. And that although we cannot see the capillaries themselves.

From the inquiry in the preceding chapter, it would appear that the result to the eyes, which has just been described, arose from implication of the roots of the cephalic vaso-motor branches of the sympathetic nerve in the morbid state of the spinal cord in the cervico-dorsal region, brought on by the concussion or injury. The cases of spinal concussion or injury in which no symptoms of affection of the sight have presented themselves, had most probably been cases in which

the spinal cord had suffered only below the cervico-
dorsal region, and in which, therefore, the roots of the
vaso-motor nerves belonging to the arteries of the optic
nervous apparatus, had not been implicated.

CHAPTER XVI.

ANALYSIS OF THE CONDITIONS ON WHICH THE MOVE-
MENTS OF THE PUPIL DEPEND, AND BY WHICH
THEY ARE REGULATED.

THE conditions on which the movements of the pupil
depend, and by which they are regulated, are so com-
plex, that it will be useful to make an analysis of
them, preparatory to an examination in further detail
of the state in which the pupils were found in the
cases under consideration, and an inquiry into how
far they are affected, in common with the arteries of
the optic nervous apparatus, in injury of the spinal
cord.

The state of relaxation of the iris is that in which
the pupil is neither much contracted nor much dilated;—a medium state into which the pupil falls some
time after death, in consequence of an elasticity
the tissue of the iris is endowed with, and to which,
during life, it has, in consequence of the same elasticity,
a constant tendency to return, when the dilating or
contracting muscular force ceases to act.

During life, contraction of the pupil to a smaller
than the medium size, and dilatation of it to a larger,

must consequently be viewed as manifestations of an active state. Contraction of the pupil to a smaller than the medium size being a manifestation of the action of the *sphincter pupillæ*, or circular muscular fibres; and dilatation of the pupil to a larger than the medium size, being a manifestation of the action of the *dilator pupillæ* or radiating muscular fibres of the iris. The contraction of either of these sets of muscular fibres having ceased, it is, as just stated, the elasticity of the iris which brings the pupil back to its medium diameter.*

ACTION OF THE SPHINCTER PUPILLÆ MUSCLE.

The contractile power of the sphincter pupillæ is under the government of fibrils from the oculo-motor nerve; whilst the contractile power of the dilator pupillæ is under the government of fibrils which the muscle derives from the sympathetic in the neck, in common, as we have seen, with the circular muscular fibres of the arteries.

The stimulus which ordinarily calls forth the action of the sphincter pupillæ muscle, and so excites contraction of the pupil, is the light. As is generally acknowledged, it is not by any direct action on the iris, that light excites the action of the sphincter, and thus calls forth contraction of the pupil. It is the impression of the light on the retina which excites, by reflex action,

---

* See the Author's paper "On the Motions of the Pupil," in the *Edinburgh Medical and Surgical Journal,* vol. xli., pp., 40—42. 1834.

the branch of the oculo-motor nerve, which supplies
nervous influence to the sphincter pupillæ; the centre
of this reflex action being the anterior pair of the
corpora quadrigemina in the brain. Hence an essential condition for the action of light on the pupil is
that the path from the retina, through the optic nerve
to the anterior pair of the corpora quadrigemina, and
thence back through the oculo-motor nerve to the iris,
be not interrupted.

If in an animal—a rabbit for example—the optic
nerves within the cranium be exposed, and one of them
divided, it is found that the brightest light produces no
effect on the pupil of the side operated on, though the
pupil of the uninjured side remains obedient to light as
before. When one of the anterior pair of the corpora
quadrigemina is removed, blindness of the opposite eye
is induced, and its pupil ceases to contract under the
influence of light. In cases of complete paralysis of
the oculo-motor nerve, whether from disease in man or
section by way of experiment in an animal, the impression of light on the retina has no influence in exciting
contraction of the pupil.

When the optic nerve has been divided, irritation of
the end of the segment in connection with the brain
excites contraction of the pupil, but irritation of the
end of the segment in connection with the eyeball does
not. On the other hand, if the oculo-motor nerve be
cut, it is the end of the segment in connection with the
eyeball,—irritation of which excites contraction of the
pupil,—not the end of the segment in connection with

the brain. Simple irritation of the corpora quadrigemina is found to excite contraction of the pupils.

Contraction of the pupil in consequence of irritation
of the end of the segment of a divided optic nerve
which is in connection with the brain, is owing to reflex
nervous action from the optic nerve through the corpora
quadrigemina to the oculo-motor nerve. Whereas, the
contraction of the pupil, excited by irritation of the end
of the segment of a divided oculo-motor nerve, which is
in connection with the eyeball, is an example of direct
nervous action. As is also contraction of the pupils excited by simple irritation of the corpora quadrigemina.

When the oculo-motor nerve is paralysed, there may
be dilatation of the pupil to a somewhat greater than a
medium degree, from paralysis of the sphincter pupillæ
permitting a more or less unrestrained action of the
dilator muscle. In such a case, the pupil is still capable
of becoming more dilated by the increased action of the
dilator, but on the relaxation of that muscle it comes
back again nearly to a middle state by virtue of the
elasticity of the iris. Thus, in a case, at present under
my care, of ptosis and inability to turn the eye inwards,
upwards or downwards, from paralysis of the third
nerve, when the eye, after being darkened, was suddenly opened to the light, the pupil, which had become
slightly more dilated by the action of the dilator
muscle called forth by the darkness, was brought to
the middle state by the elasticity of the iris. After
which, although still exposed to the light, it fell back
again to the slightly dilated state in which it was before;

the antagonism of the elasticity of the iris being only nearly, but not quite, sufficient to restrain the passive action of the radiating fibres.

In this case of complete paralysis of the oculo-motor nerve, the pupil still admitted of being dilated to the full by the action of atropia.* The pupil also admitted of being contracted to the size of a pin's head by the action of Calabar bean dropped into the eye in the form of a solution of the extract. After the influence of the atropia passed off, the pupil regained its previous medium size. After the influence of the Calabar bean passed off, the pupil also fell back into its previous medium size. From this it is evident that it is the medium size which represents the passive state of the pupil—the state of relaxation of the iris—and not the full dilatation, as has been so generally supposed.

When, on the other hand, the dilator pupillæ is paralysed by section of the sympathetic in the neck in dogs and rabbits, the unrestrained action of the sphincter does not take place to its fullest extent. The pupil, though smaller than that of the uninjured side in the shade, has been found by Professor Budge to be still capable of increased contraction when the eye was exposed to the light.

In like manner it has been found that after division of the sympathetic in the neck the pupil admits of

* This action of belladonna in dilating the pupil in a greatly increased degree, as observed in a case of paralysis of the third nerve, under the care of Dr. Latham and Dr. Burrows at St. Bartholomew's Hospital, is mentioned by the late Dr. Baly, in his "Translation of Müller's Physiology," 1839.

N

being contracted by the action of Calabar bean to a still greater degree than that to which it became contracted as a result of the section.

When the dilator pupillæ or radiating muscular fibres of the iris are paralysed by section of the sympathetic in the neck, the tendency to contraction of the pupil beyond the middle state is greater than is merely the result of the unrestrained action of the sphincter pupillæ. This, I have no doubt, is owing to congestion of the iris, which, as well as congestion of the conjunctiva and other parts of the eye and side of the head, takes place, as we have seen, after section of the sympathetic in the neck. In his experiments, Dr. Brown-Sequard has noted *discoloration of the iris* as one of the effects of the section, discoloration indicating, as is well known, congestion. It is to be remembered that when the iris is in a state of inflammatory congestion the pupil tends to contract, and yields incompletely to atropia.

When the sensibility of the retina is morbidly augmented the pupil is excited to contraction by the impression of a lesser degree of light. When, on the contrary, the sensibility of the retina is morbidly impaired, the pupil remains dilated, notwithstanding the impression of a greater degree of light. It is, however, to be observed that the susceptibility of the retina to the impression of light on which movements of the pupil supervene, is not always in proportion to the visual energy of the retina. This latter may be much impaired, and yet the impression of the light will excite the movements of the pupil in a normal manner.

ACTION OF THE DILATOR PUPILLÆ MUSCLE.

A contracted state of the pupil, from paralysis of its dilator muscle, permitting, to a certain extent, unrestrained action of its sphincter, was, we have seen, one of the effects of the experiment of dividing the sympathetic nerve in the neck of a dog, observed by Pourfour du Petit. All subsequent experimenters bear testimony to the fact, not only in the case of the dog, but also in the case of other animals experimented on, such as the cat and rabbit.

A similar state of the pupil having been found to be an effect of injury of the spinal cord between the sixth cervical and fourth dorsal vertebræ, or even down to the ninth or tenth dorsal vertebra, it has been concluded that the region of the spinal cord comprehended between those limits is the spinal centre of the sympathetic nerve fibrils, on which the contraction of the dilator pupillæ muscle or radiating muscular fibres of the iris depends.

In exemplification of this, the following cases are adduced :—

In a man who had suffered an injury of the spine, and whose pupils were contracted before death, there was found, as related by Sir Benjamin Brodie,[*] a small

---

[*] Pathological and Surgical Observations, relating to Injuries of the Spinal Cord, in the "Medico-Chirurgical Transactions," vol. xx., p. 149, 1837.

extravasation of blood in the centre of the spinal cord, opposite the fifth and sixth cervical vertebræ.

In a case of injury of the spine, observed by Mr. G. Busk, in which death took place on the seventh day, and in which the pupils were contracted latterly, though natural at first,* the interior of the spinal cord was found, on dissection, converted into a red pulpy matter for the extent of an inch and a half; the middle of this space being opposite the root of the fifth cervical nerve.

It may be further mentioned that Dr. Brown-Sequard † has recently dwelt on the frequent occurrence of contracted pupil in cases of hemiplegia from injury of the spine in the cervico-dorsal region.

It is thus seen that the part of the sympathetic nerve which governs the contractions of the radiating muscular fibres of the iris is the very part which governs the contractions of the circular muscular fibres of the walls of the internal carotid and vertebral arteries and their ramifications. Hence the coincident dilatation of the arteries and contraction of the pupil, as effects of section of the sympathetic in the neck, or destruction of its roots in the spinal cord,—the dilatation of the arteries being owing to paralysis of their walls; the contraction of the pupil being owing to the unrestrained action of the sphincter pupillæ (which is governed by the oculo-motor nerve) in consequence of the paralysis of the dilator pupillæ.

---

* Dr. W. Budd, on the Pathology of the Spinal Cord, in the " Medico-Chirurgical Transactions," vol. xxii., p. 179, 1839.

† *Ut supra*, p. 162.

After section of the sympathetic nerve in the neck, irritation of the upper segment of the divided nerve by galvanism, we have seen, p. 104, excites contraction of the dilator pupillæ, and thereby produces dilatation of the pupil as well as constriction of the arteries,—the latter as manifested by diminution of the vascular fulness and heat of the corresponding side which had supervened on the section of the nerve.

According to the experiments of Budge, the communicating branch between the hypoglossal nerve and the superior cervical ganglion of the sympathetic, conveys, as has been previously stated, from the medulla oblongata to that part of the sympathetic which governs the contractions of the radiating muscular fibres of the iris, a set of motor fibrils in addition to those which the sympathetic in the neck below the superior cervical ganglion down to the inferior, receives. The part of the medulla oblongata indicated is, therefore, viewed by Budge as an additional centre of the iridal sympathetic, and named by him *superior*, in contradistinction to that part of the cord between the sixth cervical and fourth dorsal, which part he calls the inferior spinal centre of the iridal sympathetic.

The condition under which the action of the dilator pupillæ muscle is ordinarily called forth, and dilatation of the pupil thereby excited, is an absence of the stimulus of light on the retina. If, then, the negative state of excitement into which the retina falls, in the absence of the impression of light on it, be the condition which calls forth the activity of the sympathetic fibrils distributed to the dilator pupillæ muscle, we have to inquire into the mechanism by which the effect is brought about.

Of this mechanism, we have the materials in the irido-spinal centres of the sympathetic in the cord, in the excitory or centripetal fibrils, and in the motor or centrifugal fibrils of the sympathetic, only the exact nature of the connection between the retina and the excitory fibrils which in that case would exist, has not been made out,—whether the negative impression of darkness is conveyed to the brain through the optic or fifth nerve, and thence to the irido-spinal centre,—or whether it is conveyed in a more direct manner to the irido-spinal centre by means of a connection between the retina and excitory fibrils of the iridal sympathetic, remains a question.

The extreme dilatation of the pupils in some cases of amaurosis must be owing to spasmodic contraction of the dilator pupillæ, as there is no reason to presuppose paralysis of the sphincter, even if such paralysis were a condition for unrestrained action of the dilator to so great a degree as to cause extreme dilatation of the pupil, which it is not. The spasmodic contraction of the radiating fibres may be supposed to be owing to reflex action of the nerves on which the contractile power of the radiating fibres depends, called forth by the insensible state of the retina in a manner analogous to that in which the absence of the excitement by light on the healthy retina calls forth dilatation of the pupil, whilst in consequence of the same insensible state of the retina, reflex action of the nerves, on which the contractile power of the circular fibres depends, is no longer called forth by the stimulus of light.

CONSENSUAL MOVEMENTS OF THE PUPILS.

Under this head we have to notice the movements of the two pupils in concert with each other, and the movements of the pupils in concert with the movements made in directing the eyeballs to different distances.

When one eye only is exposed to the light, both pupils are nevertheless excited to contract, though the degree of contraction they undergo is less than when both retinæ receive the impression of light. This consensus between the sphincter muscles of the two irides depends on the connexion with the corpora quadrigemina which the two optic nerves have in common, as demonstrated by the experiments of Flourens, who found that irritation of the corpora quadrigemina on the right side called forth contraction of both pupils, though the contraction of the left pupil was in a less degree.

There does not appear to be any common reflex centre for the nerves of the two dilators of the pupil, such as is here shown to exist for the nerves of the two sphincters.

In some cases of amaurotic blindness the motions of the pupil under the impression of light are natural.

This would show that the retina and optic nerve, though incapable of receiving and transmitting visual impressions, are still susceptible of the stimulus of light, and that the path for reflex nervous action through the corpora quadrigemina to the iridal portion of the oculo-motor nerve, is entire.

In most cases the motions of the pupil are either imperfectly or not at all excited, but it not unfrequently happens in cases of uniocular amaurosis that the pupil of the blind eye, though it remains in a state of fixed dilatation so long as the sound eye is covered, contracts in concert as soon as the latter is exposed to the light, and its pupil thereby excited to contract. In such a case the path of the nervous influence was along the optic nerve of the sound eye to the corpora quadrigemina, and thence through the iridal portion of the oculo-motor nerve to the sphincter pupillæ of the blind eye, as well as to that of the sound eye.

Dilatation of the pupils with adjustment for distant sight is involuntarily determined by the voluntary act of directing the eyes, with their visual axes or lines parallel, to a distant object. On the other hand, contraction of the pupils with adjustment for near sight is involuntarily determined by the voluntary act of converging the visual lines of the two eyeballs on a near object. These movements of the pupils take place independently of the strength of the impression of light on the retina. In the first instance the dilator muscles of the pupil act in concert with the two external recti if we look straight forward, or with the external rectus of one side and the internal rectus of the other side if we look sideways. In the second instance the sphincter muscles of the pupil act in concert with the two internal recti muscles. We sometimes see in cases of amaurosis that the pupils, though unaffected by light, will contract when the eyeballs are converged. During sleep, in which the eyes are in general turned inwards and upwards, the pupils are contracted. This has even been observed in amaurotic persons whose pupils were dilated in the daylight.

It is worthy of remark that, while it is the third nerve which governs the action of the sphincters in contracting the two pupils on the one hand, and the action of the internal recti in producing the convergence, consensual therewith, of the two eyeballs on the other ; it is the sympathetic which, at the same time that it governs the consensual dilatations of the two pupils, appears to have some share in determining the action of the external recti, which is consensual therewith. The fact that, after section of the sympathetic in the neck, the eyeball is more or less turned inwards at the same time that the pupil is contracted somewhat, gives support to this idea. For the inclination of the eyeball inwards must be owing to impairment of the power of the sixth nerve, arising in some manner from the section of the sympathetic. The sixth nerve, it is known, is joined, in traversing the cavernous sinus on the outer side of the internal carotid, by two or three small filaments from the plexus of the sympathetic accompanying that artery. Indeed it was this connection with the sixth nerve which gave rise to the old notion that the sympathetic had its first origin from that nerve.

CHAPTER XVII.

ACTION OF MYDRIATICS AND MYOTICS.

In continuation of the study of the movements of the pupil, the important question as to the mode of action of atropia and Calabar bean on the iris and on the apparatuses of adjustment, now claims consideration.

As allied to this question, the curious facts that atropia causes constriction of arteries, and Calabar bean constriction of veins, discovered by me, will be noticed.

ACTION OF ATROPIA ON THE PUPIL.

Considering that the state of relaxation of the iris is that in which the pupil is neither much contracted nor much dilated, and that contraction and dilatation of the pupil are manifestations of active states—the former of the circular, the latter of the radiating muscular fibres of the iris,—it is to be inferred that the action of atropia in calling forth dilatation of the pupil, consists in stimulating to contraction the radiating fibres.

In explanation of the mode of action of mydriatics,

it is not necessary to assume that the sphincter pupillæ or circular muscular fibres become paralysed. In paralysis of the sphincter, the pupil falls, as we have shown, into the middle state by virtue of the elasticity of the iris, or, at the most, somewhat dilated by the unrestrained action of the radiating fibres, though not to the full extent of their action. This is shown by the fact, that in complete paralysis of the oculo-motor nerve, the pupil admits of being very much more dilated by the action of atropia, and also by the fact that after section of the oculo-motor full dilatation of the pupil does not take place, but that this effect is produced by belladonna as usual.

After section of the sympathetic in the neck, atropia still exerts its dilating influence on the pupil, though less readily. In reference to this it is to be remarked, that when the iris is in a state of congestion or inflammation, the pupil tends to contract, and yields incompletely to the influence of belladonna. Now, that the iris is in a state of congestion after section of the sympathetic in the neck, as well as the conjunctiva and other parts of the side of the head, has been found by Dr. Brown-Sequard. Hence there is, on that account, a greater tendency to contraction of the pupil than is merely the result of the unrestrained action of the sphincter pupillæ, permitted by the paralysis of the dilator. If belladonna acted merely by paralysing the sphincter, we could not have such a result as this in any degree, seeing that the dilator pupillæ, already so completely paralysed by section of its nerve, would not be in a condition to act spontaneously on the supposed cessation of the antagonism of the sphincter. In thus exciting the radiating muscular fibres of the iris, the atropia must exert a direct local action on them or on the sympathetic fibrils in the iris. That a solution of atropia dropped into the eye makes its way by endosmose into the interior, is shown by the following experiment :—The aqueous humour of a rabbit, into the eye of which a solution of atropia had been repeatedly dropped, being evacuated, collected, and dropped into the eye of a dog, was found by Dr. Von Graefe to produce dilatation of the pupil in the latter animal.

In arguing that atropia dilates the pupil by exciting to contraction the dilator pupillæ muscle, I have always meant *by muscle* the proper muscular fibres, together with the fibrils of the iridal sympathetic distributed among them.

What has here been said of the action of atropia on the pupil is applicable to the other agents known as mydriatics, such as hyosciamus and stramonium.*

ACTION OF CALABAR BEAN ON THE PUPIL.

Under the action of irritating applications to the conjunctiva the pupil becomes contracted for the time, but such action on the pupil is not the specific counter-

---

* I have elsewhere (" British and Foreign Medical Review," the " Medical Times and Gazette," and my " Ophthalmic Medicine and Surgery," 3rd edition) examined in detail the various hypotheses which have been entertained on the mode of action of atropia in dilating the pupil.

part of that exerted by belladonna. The only known agent which specifically excites contraction of the pupil, as belladonna causes dilatation of it, is CALABAR BEAN,* and for this discovery, medical science is indebted to Dr. Thomas R. Fraser of the University of Edinburgh.† Dr. Fraser at the same time showed that, along with the contraction of the pupil, adjustment of the eye for near sight was excited,‡ just as along with dilatation of the pupil by atropia an extreme state of adjustment of the eye for distant sight is, in an analogous manner, excited. Furthermore, as dilatation of the pupil is a symptom of the general action of belladonna on the system, so contraction of the pupil was also found by Dr. Fraser to be an effect of the general action of the Calabar bean on the system.

Contraction of the pupil by Calabar bean, which is to a greater degree than that excited by the strongest action of light, does not continue so long as the dilatation of the pupil by atropia.

In the medium dilatation of the pupil from paralysis of the oculo-motor nerve, the Calabar bean we have seen (p. 177) excites contraction of the pupil, no doubt, by direct action on the sphincter. For solution of Calabar bean like that of atropia, locally applied, will

---

* Physostigma venenosum.

† On the Ordeal Bean of Calabar, " Edinburgh Medical Journal," July, August and September, 1863.

‡ I am happy to have here the opportunity of repairing an oversight at page 452 of my " Ophthalmic Medicine and Surgery," by which I did less than justice to Dr. Fraser, in my anxiety to do justice to all.

make its way by diffusion into the interior of the eye. It is easy, says Dr. Fraser, to prove the presence of the extract within the eyeball after its topical application, by removing the aqueous humour and placing it on the conjunctiva of another animal, when the usual effects of Calabar bean on the pupil will be produced.

Seeing that the state of relaxation of the iris is that in which the pupil is neither much contracted nor much dilated; and that contraction and dilatation of the pupil are manifestations of an active state,—the former of the circular, the latter of the radiating muscular fibres of the iris,—it is to be inferred that the mode of action of the Calabar bean, in producing contraction of the pupil, consists in exciting contraction of the sphincter pupillæ, or circular muscular fibres, either by direct action on them, or through the medium of the oculo-motor nerve which governs their contractions.

In explanation of the mode of action of the Calabar bean, it is not necessary to assume that the radiating muscular fibres of the iris become paralysed. It has, indeed, been found that the Calabar bean always excited greater contraction of the pupil in dogs and rabbits after section of the sympathetic in the neck; but I have already remarked that, under such circumstances, there is, in consequence of congestion of the iris, an increased tendency to contraction of the pupil. Besides, irritation of the sympathetic still excited the dilatation of the pupil, though under the influence of Calabar bean.

ACTION OF ATROPIA AND CALABAR BEAN ON THE ADJUSTMENT OF THE EYE.

As belladonna, besides exciting dilatation of the pupil, excites an extreme state of adjustment of the eye for distant sight, so the Calabar bean, besides exciting contraction of the pupil, excites an extreme state of adjustment of the eye for near sight, as was first discovered and pointed out by Dr. Fraser.

Whilst dilatation of the pupil by atropia is evidently owing to contraction of the dilator pupillæ or radiating fibres of the iris, the coincident adjustment for distant vision is as evidently owing to contraction of muscular fibres subservient to that purpose.* So also, whilst contraction of the pupil by Calabar bean is evidently owing to contraction of the sphincter pupillæ, or circular fibres of the iris, the coincident adjustment for near vision is as evidently owing to contraction of the muscular fibres subservient to that action.

ACTION OF ATROPIA IN CAUSING CONSTRICTION OF ARTERIES.

In the course of my researches on the state of the blood and the blood-vessels in inflammation, I found that the application to the frog's web of a solution of the sulphate of atropia (four grains to the ounce of water) was followed by constriction of the arteries of the part in about the same time that the dilatation of

---

* For the arguments on this point, see my " Ophthalmic Medicine and Surgery," 3d edition, pp. 459-465, 1865.

the pupil supervenes on the dropping into the eye of
the same solution. The application of a stronger solution of atropia was followed by a still greater degree of
constriction of the arteries.

As the constriction of arteries is owing to contraction
of the circular fibres composing their muscular coat, it
follows that the *modus operandi* of the atropia in this
case must be to excite the contraction of those fibres.

Contraction of the circular muscular fibres of the
arteries manifested itself by the increasing thickness of
the wall, in proportion as the caliber of the vessels
became constricted.

In first recording the observation which I have now
mentioned in my essay " On the State of the Blood and
the Blood-Vessels in Inflammation," published in 1850,
I identified the action of atropia in exciting contraction
of the circular fibres of the muscular coat of the
arteries, with its action in causing dilatation of the
pupil.

A solution of sulphate of atropia (gr. ij to water ʒj)
was freely applied to the ear of a white rabbit over the
artery which runs up the middle. On examination in
about twenty minutes after, the artery was found constricted, and the size of the veins diminished. It is
here to be remembered that the artery in the natural
state may be observed to become spontaneously constricted and then dilated at short intervals, though not
rhythmically, and that the vascular injection of the ear
is reduced or augmented accordingly, whilst the heat of
the ear is in proportion to the fulness of the vessels.

The effect of the atropia on the artery was not to excite contraction of its walls and constriction of its calibre in the greatest degree, but only to keep up a more continuous state of constriction than natural—the artery being observed to be still capable occasionally, though less frequently, of becoming constricted to closure of its lumen.

The veins were not directly constricted by the action of the atropia, though they became so indirectly. Thus while the artery was constricted there was a less afflux of blood into the veins, and they accordingly became small.

In saying that the *vessels* of the rabbit's ear are contracted by the action of atropia, Dr. Brown-Sequard* overlooked the mechanism I have described, and attributed, by implication, to atropia an action on the walls of the capillaries and veins which it does not possess.

I have found that atropia exerts its constricting action on the arteries of a frog on the side on which the ischiatic nerve has been cut, as well as on the side on which the nerve was still entire.

It is to be observed that the slowly progressing and moderate constriction of the arteries, due to the action of atropia, must be distinguished from the sudden constriction to closure, even, of the vessels which, as before mentioned, is now and then seen to take place

---

* In the sequel to his paper on " Section and Galvanization of the Sympathethic Nerve in the Neck," in the *Gazette Médicale de Paris* for 1854, entitled " Experiences prouvant qu'un simple afflux de sang à la tête peut être suivi d'effets semblables à ceux de la section du nerf grand sympathetique."

independently of the application of atropia. It is important to remark that an artery while under the influence of atropia may, from some occasional cause, become constricted to closure in the manner above mentioned, and speedily become open again. The dilatation, however, which succeeds this transient closure is only back to the diminished width of the artery, already caused by the atropia which had, so far, excited the circular muscular fibres of the vessel. An analogous phenomenon may be observed in the case of the pupil. A pupil not yet fully under the influence of atropia may be observed to become transiently still more dilated by darkening the eye, but on exposure to the previous degree of light to come back to the less degree of dilatation due to the, as yet, incomplete action of the atropia.

In the contraction of the circular muscular fibres of the arteries, as manifested by the increasing thickness of their walls, and the resulting constriction of the calibre of these vessels, excited by the application of atropia, we have an ocular demonstration that that mydriatic is a specific stimulant of the circular muscular fibres of the arteries, as it is of the radiating muscular fibres of the iris.

Congestion of the conjunctiva and eyeball is sometimes induced by the action of atropia, but when its use is discontinued the congestion subsides. Taught by my microscopical observation of the action of atropia applied to the frog's web in causing moderate constriction of the small arteries, whereby congestion of blood corpuscles is permitted gradually to supervene in the corresponding capillaries and venous radicles, I view the congestion of the eye in such a case as the effect of constriction of its small arteries induced by the action of the atropia. Seeing that constriction of the small arteries of a part may cause congestion of blood corpuscles in the capillaries and venous radicles, to which they lead, and seeing that the fundamental morbid condition of the eye in glaucoma is great venous congestion, we can understand how it is that belladonna or atropia applied to the eye in glaucoma aggravates the symptoms by increasing the constriction of the small arteries of the choroid and retina, and so determining an augmentation of the congestion in the capillaries and veins. When the glaucomatous state is not much developed, the eye slowly recovers from the injurious operation of the belladonna. This may not, however, prove to be the case in an advanced stage of the disease. In a stage in which the sight is already much deteriorated, a single application of belladonna to the eye is likely to deteriorate it still more, and that irrecoverably.*

ACTION OF CALABAR BEAN IN CAUSING CON-STRICTION OF VEINS.

Entertaining the opinion that the action of Calabar bean in producing contraction of the pupil consists in exciting to contraction the circular muscular fibres of

---

* See the last edition of my work on " Ophthalmic Medicine and Surgery," p. 143, 1865.

the iris, I became curious to ascertain if it would have any effect on the walls of the veins or other blood-vessels, such as I had long before ascertained belladonna to exert on the arteries. For the experiments on the contractility of the walls of the blood-vessels, I have found anæmic frogs best adapted, both because variations in the width of their calibre are more easily observed, and because frogs in this condition appear to be more susceptible to impressions. The web of a frog, then, being disposed under the microscope, a strong solution of the extract of Calabar bean was applied to both its surfaces. The result was that the arteries appeared to be transiently affected, namely, first constricted, and then dilated somewhat. But this I consider to be not a specific effect, but only such as I have observed to be produced by any irritating application to the web. No effect was observed on the capillaries, but on the veins an effect was observed which I consider to be a manifestation of specific action.

This effect on the veins consisted in constriction of their calibre to as great a degree as is ever seen to take place in the veins of the web of the frog under any circumstances. The degree and form of constriction which presented itself in a few minutes after the application of the solution of the Calabar bean to the web was this: The vessel became diminished in width, though in a very slight degree, perhaps to about one-seventh of its diameter, and here and there constrictions appeared as if a thread had been tied round the vessel at intervals, so that it had a somewhat varicose appearance. The veins in the web of the frog, it is to be remembered, have only a few circular muscular fibres in the composition of their walls here and there; hence the appearance of constrictions here and there only, in an increased degree over what may under ordinary circumstances be observed.

The web of a frog under the microscope was subjected to the influence of atropia, when the arteries only were seen to become constricted, but not the veins. Calabar bean solution* was next applied, when the veins were seen to become constricted, the arteries remaining constricted as before. There was thus presented to view both arteries and veins in a state of constriction at the same time; the arteries by the action of atropia, the veins by the action of Calabar bean.

I have found Calabar bean to exert its constricting action on the veins of the web of the leg of a frog in which the ischiatic nerve has been cut, as well as on the veins of the web of the leg of a frog in which the ischiatic nerve was still entire.

Having shaved the fur from the ear of a rabbit, and with fine sand-paper rubbed the epidermis thin over a vein, I waited a few minutes and then applied the solution of the extract of Calabar bean to the place. Watching the part I very soon saw the veins become much constricted, and at some places varicose. On account of the varying width of the artery, and the

---

* A piece of the extract, about the size of a grain of wheat, reduced to the state of an emulsion with three or four drops of water, was the solution I have used in my experiments.

consequent variation in the quantity of blood transmitted to the veins, the exact degree of constriction of
the veins due to the action of the Calabar bean did not
last long. But the vein continued for a considerable
time to be appreciably narrower at the place where the
Calabar bean was applied than it was either above or
below. When the artery became much dilated, and a
free afflux of blood to the veins was permitted, the constriction became less evident than it was when the
veins were in a medium degree of distension.

In several patients the veins of the ocular conjunctiva were examined with a magnifying glass, and their
breadth noted as far as was possible. Calabar bean in
the form of the extract dissolved in water was then
inserted into the eye, and the veins were examined at
intervals of a few minutes to see if they underwent any
change in breadth. The result was, that an appreciable constriction of their calibre was observed to take
place *pari passu* with the contraction of the pupil.

On applying Calabar bean to the web of the bat's
wing under the microscope, I observed that the rhythmical contractions of the veins were rendered stronger,
though not more frequent—the frequency both before
and after the application being ten in a minute.

In the contraction of the circular muscular fibres of
the veins and the resulting constriction of the calibre
of those vessels excited by Calabar bean, we have thus
an ocular demonstration that that agent is a specific
stimulant of the circular muscular fibres of the veins,
as it is of the sphincter pupillæ or circular muscular

fibres of the iris. The action of Calabar bean in causing constriction of the veins indicates that the muscular coat of those vessels is governed by a motor-nervous influence belonging to a category similar to that to which the influence of the branch of the oculo-motor governing the contraction of the sphincter pupillæ or circular muscular fibres of the iris belongs —a motor nervous influence opposed to the sympathetic nervous influence which governs the contractions of the circular muscular fibres of the small arteries and those of the dilator pupillæ, or radiating muscular fibres of the iris.

CHAPTER XVIII.

ACCORDING to the investigations in the preceding chapter, actual injury of the roots of the sympathetic nerve in the cervico-dorsal region of the spinal cord would have for effect paralysis of the dilator pupillæ or radiating muscular fibres of the iris, and of the circular muscular fibres of the arteries, with a consequent tendency to contraction of the pupil from the unrestrained action of the sphincter pupillæ, and tendency to dilatation of the arteries from distension by the blood.

Irritation of the roots of the sympathetic in the cervico-dorsal region of the spinal cord, from morbid action going on there, would, on the contrary, have for effect contraction of the radiating muscular fibres of the iris, and of the circular muscular fibres of the arteries, with consequent dilatation of the pupil in the one case, and constriction of the arteries in the other.

These effects are, as above stated at p. 161, re-

markably exemplified in cases of locomotor ataxy, in
which it has been observed that during the intervals
between paroxysms of pain the pupil was contracted,
and the conjunctiva injected; whilst during a paroxysm
the pupil became more or less dilated, and the conjunctiva free from injection. In other words, during
a paroxysm of pain, the roots of the sympathetic in
the cervico-dorsal region of the spinal cord are, in common with the roots of sensitive nerves, in a state of
irritation, the consequence of which is the excitement
to contraction of the radiating muscular fibres of the
iris, and of the circular muscular fibres of the arteries,
with dilatation of the pupil in the one case, and constriction of the arteries in the other.

In cases of locomotor ataxy, it has been observed by
Dr. Auguste Voisin * that if any portion of the skin of
the upper extremities, or of the trunk—or, better still,
of the lower extremities, be pinched, pricked, gal-
vanised, or otherwise irritated, dilatation of the pupils
immediately follows. The explanation of this, which
suggests itself, is that the iridal sympathetic partici-
pates in the excitement into which the diseased spinal
cord, in which it has its roots, is thrown in consequence
of the impression transmitted to it from the irritated
skin by the sensitive nerves; and the stimulus to contract is thereby communicated to the dilator pupillæ.
The dilatation of the pupil, produced by pinching the
skin, would thus appear to be analogous, in its proxi-

* Gazette Hebdomadaire de Médecine et de Chirurgie.   Paris,
1864.   No. 38, p. 633.

mate cause, to that which occurs during a paroxysm of pain.

With the dilatation of the pupil as excited in the manner just described, may be collated the constriction of the arteries observed in the following experiment recorded by Dr. Nothnagel of Königsberg.* Having in a rabbit exposed the pia-mater of the brain by trepanning the skull and removing the dura-mater at the place, Dr. N. irritated the skin of the leg over the crural nerve by galvanism, whereupon a distinct constriction of the arteries of the pia-mater was observed to take place. On the cessation of the irritation, the constriction gradually gave way to dilatation.

In our cases of concussion of the spine, there was no actual paralysis of the dilator pupillæ; nor, on the other hand, increased action of it in any marked degree. This might have been owing to the superior spinal centre of the iridal sympathetic not being implicated in the mischief going on in the spinal cord lower down, where the iridal sympathetic has its inferior centre in common with the sympathetic which governs the contractions of the walls of the arteries of the head. (See pp. 72—103-4.)

Though in the majority of our cases a tendency to contraction of the pupil manifested itself, there was

---

* Virchow's Archiv, Bd. xl., p. 210. In his paper (which is entitled " Die vasomotorischen Nerven der Gehirngefässe ") Dr. Nothnagel relates some experiments on the rabbit, from which it would appear that the carotid plexus receives additional vaso-motor nerves from a source above the superior cervical ganglion—probably through cerebral nerves.

still observed considerable power in the dilator pupillæ. Sometimes, even, there was a tendency of the pupil to become somewhat dilated. This phenomenon might be attributed to simple irritation of the roots of the iridal sympathetic, though in explanation of it, it becomes necessary to take into account the degree of sensibility still possessed by the retina. Whilst contraction of the pupil in some of our cases was evidently a reflex of the morbid irritability of the retina, arising from the congested state of the optic nervous apparatus, —the tendency to dilatation observed in other cases was no doubt in part owing to the impaired sensibility of the retina.

The tendency to dilatation of the pupil may have been indirectly owing to the impaired power of maintaining the near adjustment of the eyes, and keeping them converged on the object looked at, though more likely it was an effect of the overwhelming action of the radiating muscular fibres of the iris, together with the muscular fibres which are subservient to adjustment of the eyes for extreme distance—they being, as well as the dilator pupillæ, under the government of the sympathetic nerve.

Atropia produced dilatation of the pupil in our cases as usual. This effect, however, would have been obtained even if the nerves of the radiating muscular fibres had been quite paralysed. For atropia, we have seen, is capable of exerting a direct local action on the dilator pupillæ, in cases in which the sympathetic nerve in the neck has been cut.

CHAPTER XIX.

OF THE FAILURE OF HEARING IN THE CASES UNDER
CONSIDERATION, AND ITS DEPENDENCE ON THE
INJURY OR DISEASE OF THE SPINAL CORD.

In some of our cases the sense of hearing was affected. The patients complained of impaired perceptive power, with hyperacousia or excessive sensibility to objective sounds, and of hearing subjective sounds, or noises in the head.

From the carotid plexus of the sympathetic, some filaments pass into the tympanum, through the petrous bone, and join the nervous anastomoses of Jacobson—an anastomoses among the said filaments,—filaments of the Vidian nerve—and filaments of the glosso-pharyngeal,—which lies on the inner wall of the tympanum and gives off branches which accompany small arteries to the membranes of the two fenestræ.

Filaments from the inferior ganglia of the sympathetic in the neck accompany the vertebral artery into the cranium, forming around it a plexus. The internal auditive artery, it will be remembered, is derived from the superior cerebellar branch of the basilar, or from the basilar itself, which is formed by the inosculation of the two vertebral arteries, and receives an offshoot from the vertebral plexus of the sympathetic.

From this we can understand how the circulation in the auditory nervous apparatus may be disturbed in consequence of the affection of the spinal cord as well as the circulation in the optic nervous apparatus. In short, if the optic nervous apparatus suffers from injury of the spine, as we have seen it does, the auditory nervous apparatus also suffers from the same cause, operating in a similar manner under similar organic conditions. Thus, if the arteries of the optic nervous apparatus are accompanied by filaments from the sympathetic regulating their action, so also are the arteries of the auditory nervous apparatus. The ophthalmic artery is a branch of the internal carotid, the internal auditive artery is derived from the superior cerebellar branch of the basilar, or from the basilar itself. If, in consequence of paralysis of the sympathetic vaso-motor nerves accompanying the arteries of the optic nervous apparatus, dilatation of these vessels and increased afflux of blood to that apparatus take place, or if, in consequence of irritation of the roots, in the spinal cord, of the sympathetic vaso-motor nerves accompanying the arteries of the optic nervous apparatus, constriction of these vessels and diminished afflux of blood to that apparatus take place, the same thing, for the same reason, must hold good in the case of the auditory nervous apparatus.

PART SECOND.

PATHOLOGY OF THE AMAUROTIC FAILURE OF
SIGHT CAUSED BY RAILWAY AND OTHER
INJURIES OF THE BRAIN.

CHAPTER I.

THE failure of sight, and other cerebral symptoms, in the class of cases now to be considered, are the direct effect of the injury sustained by the brain. In cases of concussion of the spine, the cerebral symptoms, we have seen, come on in consequence of affection of the brain supervening on the inflammation of the spinal cord and its membranes, primarily excited by the accident.

In cases of injury of the head, the patient is struck down at once ; and it not unfrequently happens that, for a time, he remains completely insensible to external impressions. Recovery, which, in general, speedily takes place, is sometimes complete; while at other times the state of total insensibility is followed by one in which the sensibility is impaired, though not destroyed. The patient is not affected by ordinary impressions, but can be roused to perception. The pupils in this stage are sometimes more contracted than usual. These symptoms may wholly subside in the course of a few hours, or they may continue for three or four days. In the latter case it frequently

P

occurs that the patient regains his sensibility for a time, and then relapses into his former condition.

Where inflammation of the brain follows concussion, it sometimes happens that there is no interval of returning sense, the symptoms of concussion being gradually merged into those of inflammation. But it is also often the case that there is a considerable interval of that sort, or even a period of apparent health, before the symptoms of inflammation show themselves. Years may elapse before the patient becomes affected with any serious indications of cerebral disease.

The inflammation which succeeds to concussion, and other injuries of the head, may be more or less acute : affecting the whole contents of the cranium, and rapidly proving fatal ; or limited to some particular part of the brain, and inducing death only after a series of the most distressing symptoms, as violent headache, amaurosis, palsy, convulsions, and the like. These symptoms occur sometimes in one order, and sometimes in another.

In cases of pressure on the brain from fractured cranium with depression, or from extravasation of blood, the attending insensibility may be more or less complete. In some instances the patient lies unconscious, indeed, of what is passing around him, but capable of being roused by strong impressions on his senses ; while in other cases the loss of sense is so complete that the skin may be pinched, a lighted candle held close to the eye, and the loudest noise sounded into the ear without any evident effect. Where the cause of these symptoms is simply a fractured and depressed portion of the cranium, they show themselves immediately after the infliction of the injury; but where they depend on extravasation of blood, either accompanying fracture or independent of it, the collection of blood may form slowly, and a considerable interval of time elapse before the patient becomes insensible.

Among those who recover from fractured skull with depression, or from extravasation of blood within the cranium in consequence of an injury of the head, there are some in whom the symptoms wholly subside in the course of a few days, and others, in whom certain remains of one or more of the symptoms still exist after the lapse of many years. Such variety in restoration is remarkably the case with regard to the sentient power of the eye, the mobility of the pupil, and the activity of the muscles supplied by the third nerve.*

In the cases in which the injury of the head is of so grave a nature as has just been described, the symptoms speak for themselves, and the affection of the sight is of comparatively secondary importance. The cases in which the affection of the sight is the

* Mackenzie's "Practical Treatise on the Diseases of the Eye." Fourth edition, pp. 1044, 1057. London, 1854.

In quoting Dr. Mackenzie, I cannot refrain, as an old and attached friend, from recording my respect for his memory, and bearing testimony to the singular devotion he displayed throughout a long career of usefulness, to the advancement of Ophthalmological Science.

damage most complained of, are those which chiefly concern us here.

The cerebral disturbance with which the patient is often found to be affected immediately after the accident in cases of spinal concussion is, no doubt, owing to the brain having at the same time suffered. Such cases now come under consideration along with those in which the head is especially the seat of the injury.

In some cases the sight is found to be lost immediately after an injury of the head. In other cases, again, the injury may appear to be slight, and occasion at first no apprehension. Notwithstanding this, experience has taught us that amaurosis is still liable to come on,—more frequently, perhaps, than is generally suspected. In cases in which the sight is found to be lost immediately after an injury of the head, the blindness is, no doubt, owing either to some actual lesion of the optic nervous apparatus, such as extravasation of blood in or about the root of the optic nerve, or laceration of the brain substance itself.

CHAPTER II.

CASE I.

A MAN, A. B., came to me at University College Hospital, who had fallen on his head from a scaffold a year before. He stated that on recovering his consciousness he found the sight of the right eye gone. Since then the sight of the other has become very much impaired, and is rapidly becoming more so.

He was in a broken-down state of health.

CASE II.

J. L., aged 62, seven months ago fell downstairs, from top to bottom, on the crown of his head, and was taken up insensible. The insensibility continued, more or less complete, for about three weeks. During that time he could be roused. His scalp was hot, and he was very restless, his wife being scarcely able to keep him in bed.

On recovering his senses, he found that the sight of both eyes had become so imperfect that he was unable to recognise any person.

Before the accident his sight was good. Being originally emmetropic, and having become presbyopic, he required to use convex glasses to read with, but he was still able to see anything at a distance distinctly. Now he can see nothing distinctly at any distance. With convex glasses 10 inches focus, he can read large print, and with convex glasses 18 or 20 inches focus he can see my features a little better.

Since the accident, objects have appeared to him double sometimes, and he has occasionally seen flashes of light before the eyes. He has also become dull of hearing, and is troubled with *tinnitus aurium*.

His memory was at first impaired, but it is now rather better, though not so good as before.

When convalescent from the first effects of the injury, he was liable to become giddy, and fall down on trying to stand up. Is still weak and unable to walk steadily. Has now a slight impediment in his speech, and some difficulty in finding the proper words to express himself with.

For some time after the accident, the pupils were dilated, as I am informed by Dr. Saull, who treated the case. When the patient presented himself at the hospital the pupils were found to be obedient to the light, but so small that it was necessary to bring them under the influence of atropia in order to make an ophthalmoscopic exploration. The pupils yielded well to the action of the atropia.

On examination, the lenses were found slightly hazy, and presented cataractous striæ at their circumference.

This state, however, was not such as to interfere with a view of the fundus.

Under the ophthalmoscope, the optic discs were observed to be opaque and white in the middle, and very red, from vascular injection, in their peripheral part. Supposing the diameter of the disc divided into three, the central opaque white part occupied the middle third. This part was much cupped, and the retinal vessels were, in consequence, unusually tortuous in the first part of their course. The veins were rather distended, and the arteries constricted. The retina adjacent to the disc presented dark discoloration from pigmentous deposit.

Though the patient's general health has been benefited by tonic treatment, his sight has not improved.

CASE III.

Mr. C. D. was seated in a *coupé* carriage in front of the train when it was abruptly brought to a standstill while going at considerable speed. He was first thrown forward—dashing his head through the window in front of him — and then backwards, striking the occiput against the back of the compartment. He was much stunned at the time, and bled a good deal from several cuts of the face, caused by broken glass. After the blood was washed from his face he could see with the right eye, but the sight of the left eye was dim, and has continued so ever since.

The patient was confined to the house for some time,

and though he is now able to be about, his general health has become much broken up.

Some time after the accident both eyes were found congested. The pupils seemed to be normal, or nearly so, in size and mobility, though the left pupil did not yield readily to the influence of atropia, and, after having so yielded, did not resume its proper width for a long time.

The impairment of the sight of the left eye was to such a degree that the patient could not make out very large letters, viz., No. 20 of Jaeger's scale, at any distance, and besides this the field of vision to the left side was contracted.

Under the ophthalmoscope, the disc in both eyes was very red and ill-defined at its inner margins. The retinal veins had white lines at their sides, extending for some distance. On re-examination of the eyes some months after, the retinal veins seemed to have become more engorged and tortuous.

This case, for which I am indebted to the *late* Dr. George Rainy,* of Glasgow, appears to have been one

* It is with grief that I here insert the word *late*. When this acknowledgment was first written, Dr. George Rainy was in good health, and had before him, to all human expectation, the promise of a long career of honour and usefulness. He was the successor of the late Dr. Mackenzie as Waltonian Lecturer in the University, and as Surgeon to the Eye Infirmary. Ere many weeks, however, had passed away, George Rainy, as has been the fate of many other devoted young Physicians, succumbed—a victim to typhus fever. I may apply to him what the celebrated George Buchanan, three hundred years ago, wrote of his friend, Alexander Cockburn :—

" Si numeres annos, cecidit florente juventa ;
Si studia et mores et benefacta, senex."

of concussion of both brain and eyeballs, though the failure of sight of the left eye at the time of the accident was most probably owing to some intracranial injury.

CASE IV.

Mr. E. F., aged 45, received a blow on the back of his head by being violently jerked against the back of the carriage in which he was, in consequence of the train abruptly coming to a standstill.

He became very sick at the time, and, in a day or two, it was discovered that the sight had become dim.

The surgeon who was called in on account of this, found the optic discs the seat of vascular injection. Under treatment by leeches, mercury, and counter irritation, the sight improved greatly.

The patient being originally somewhat hypermetropic, had begun to use convex glasses a year or two before the accident. When I saw him he required 24-inch convexes to see across the street with, and 10-inch convexes to read with—the eyes being at the time under the influence of atropia, and, therefore, in the lowest state of refraction. It was with the right eye only, however, that he could really see so well.

The eyes, I found on examination, had a somewhat glaucomatous aspect, so that it is probable the sight was not quite perfect before the accident. Under the ophthalmoscope the fundus presented a dirty light reddish appearance, and the optic discs were rather white.

CHAPTER III.

INQUIRY INTO THE NATURE OF THE MORBID CONDI-
TION OF THE OPTIC NERVOUS APPARATUS INDUCED
BY THE INJURY OF THE HEAD IN THE CASES
RELATED IN THE PRECEDING CHAPTER, AND
WHICH GAVE ORIGIN TO THE DEFECT OR LOSS OF
SIGHT.

AMAUROTIC failure of sight in cases of injury of the head may be owing to actual lesion of some part of the optic nervous apparatus, or to pressure on it by extravasated blood, or to implication of it in supervening inflammation. When the sight is found to be lost immediately after the injury, there is reason, we have seen, to suspect that some lesion of the nervous substance has occurred. If the failure of sight has supervened less quickly, we may infer that extravasation of blood is the cause; whilst if some days have elapsed before the patient begins to complain that he cannot see, it is probable that inflammatory action is implicating some part of the optic nervous apparatus.

The inflammation in such a case most likely commences as a basilar meningitis.

The earlier occurrence of the amaurotic symptoms in cases of injury of the head than in cases of spinal injury, would show that they were in the former case the direct result of the injury. In cases of concussion of the eyeball, the symptoms, we shall see, were still more directly the result of the blow.

The congestion at the fundus of the eyes in the cases of injury of the head, is to be viewed as a peripheral manifestation of the meningitis, or other disturbance of the circulation within the cranium excited by the injury.

We have seen that there may be perverted and impaired visual sensibility without any evident objective morbid appearance of the eye, either external or ophthalmoscopic, so that we are left to refer the subjective symptoms to some morbid action or change in the intracranial part of the optic nervous apparatus. Morbid alterations at the fundus of the eye become, however, sooner or later visible. Such alterations, it is to be remarked, are the result of a propagation of the morbid action from the intracranial to the extracranial part of the optic nervous apparatus ; and being, therefore, secondary, were not the cause of the symptoms of perverted and impaired visual sensibility which first attracted attention.

Besides the intracranial part of the optic nervous apparatus, other parts of the brain may be affected, and that primarily. Symptoms of such affection of other parts of the brain may have preceded the visual symptoms. The nature of the case, therefore, may have been declared long before the alterations at the fundus of the eye presented themselves. This it is important to keep in view, so as to guard against laying too much weight on ophthalmoscopic appearances in the eye as a means of diagnosis of cerebral disease.

In a series of able lectures on "Optic Neuritis" as a symptom of disease of the brain and spinal cord,* Dr. T. Clifford Allbutt, of Leeds, examining the subject from a physician's point of view, shows, on the other hand, that many important changes in the optic disc precede any loss of vision, and, therefore, are not seen by ophthalmic surgeons. Having given a great deal of time and care to ophthalmoscopic explorations in cases of meningitis, Dr. Allbutt has come to the conclusion that the diagnosis in the slighter forms of tubercular meningitis more especially, which might be otherwise doubtful, is greatly aided by examining the state of the optic discs. Though in such cases no affection of the sight may have been complained of, he has discovered congestion of the optic disc and retinal vessels. Dr. A. is accordingly of opinion that the ophthalmoscope is calculated to afford the same kind of help in detecting incipient or slight degrees of tubercular meningitis that the stethoscope gives in detecting those incipient or slight degrees of ulcerative change in the lungs which, without it, are beyond certain diagnosis.

CONCUSSION OF THE EYEBALL.

A few words may here be intercalated respecting concussion of the eyeball. Amaurosis is a not unfrequent consequence of even very slight blows on the eye-

* In the *Medical Times and Gazette*, from May to August, 1868.

ball, with or without any visible injury of the organ. Blows, contusions, and wounds of the eyebrow and margin of the orbit, without any visible injury of the eyeball, may also occasion amaurosis. In such cases, the failure of sight may come on soon after the accident, being owing to intraocular extravasation of blood, or it may come on only at a later period, being the result of a slowly supervening subacute form of inflammation affecting the optic nerve and retina.

In my "Ophthalmic Medicine and Surgery," p. 59, I mention the case of an officer in the Crimea who was struck on the temple by a splinter of a shell. When he recovered himself, he found he was blind of the eye on the corresponding side. On examination of the fundus of the eye with the ophthalmoscope after his return to England, I discovered the white surface of the sclerotica exposed through a transverse laceration of both the retina and choroid.

As the frontal branch of the fifth nerve is often implicated in injury of the eyebrow, the opinion has been entertained that the amaurosis is in some manner connected therewith. Although this can scarcely be admitted as regards the amaurosis immediately excited, it is by no means unlikely that implication of the vaso-motor nerves of the eye in the injury of the fifth nerve by determining disturbance of the circulation in the eye, may prove a cause of the amaurosis which some-times comes on subsequently to the injury.

The following is a case of concussion of the eyeball, in which both the subjective symptoms and the appearances observed in the fundus under the ophthalmoscope bore a striking resemblance to those which presented themselves in our cases of failure of sight from spinal concussion in railway accidents.

CASE V.

Mr. J. B., while in Australia two years ago, was accidentally struck on the right eye with a fist. The effect was that this eye became much troubled with the appearance of flashes of light before it. He thinks that the vision of both eyes continues as distinct for both far and near objects as ever. The eyes, however, soon become tired when he reads or writes. Writing tires more than reading.

Externally the eyes appear natural. The pupil of the right eye is obedient to the light, but falls into a somewhat dilated state when the left eye is shut. The pupil of the left eye is quite natural in its movements.

Since the accident Mr. B. has given his eyes a long period of rest, and has made the voyage home to England.

Under the ophthalmoscope, I found the disc of the right eye congested and, at the temporal side, whitish. The retina all round presented some discoloration of a leaden hue. The disc of the left eye was also congested, and not well defined on the nasal side.

After five months' rest and treatment, the right eye was found notably improved. Though on closing the left eye the pupil of the right still shows a tendency to dilate, the sight continued good, and the photopsy had become less.

The asthenopic weakness still existed, though in an ameliorated degree.

Under the ophthalmoscope the optic discs were found less congested looking. It is worthy of remark that there was greater congestion remaining in the left or uninjured eye than in the right, though the retinal veins of the right eye appeared somewhat more gorged. The greater paleness of the right optic disc than that of the left is unmistakeable.

Four months subsequently the disc of the left eye was found less injected than that of the right. The disc of the latter was still reddish on the nasal side, and presented some degree of whiteness on the temporal side. The sight continued perfect, but there was no abatement of the asthenopia.

Convex glasses 36 inches focus enable the patient to read better at the distance of ten inches.

AMAUROSIS FROM TUMOURS OF THE BRAIN.

In conclusion of this chapter, I add a few remarks on the effect of tumours or other morbid states of the brain, of idiopathic origin, on the sight; premising a case of amaurosis arising therefrom, as a standard by which to illustrate and compare the cases of amaurotic failure of sight from injury of the head under consideration.

CASE VI.

Mr. W. L., aged 53, suddenly saw double two years

ago—one image above the other. A few days after the two images appeared side by side. From this time the sight of the right eye gradually became dimmer and dimmer, until it was wholly lost. The sight of the left eye has of late been failing in a similar manner, and is now so imperfect that the patient cannot see to read.

He has had shooting pains through the head from temple to temple. The strength is much impaired. He sometimes staggers in his gait and feels dizzy. Suffers, occasionally, twitchings in his legs, with a sensation of creeping.

Hearing in the right ear dull.

The conjunctiva of the lower eyelids is anæmic looking. The pupils are small and all but immovable.

Under the ophthalmoscope the choroid pigment was seen shining through the retina, and the optic disc presented the dense bluish whiteness of atrophy. The retinal veins were gorged and tortuous. This state of matters existed in both eyes, only it was in a greater degree in the right.

Though, at the commencement of the disease in this case, there was diplopia, the movements of the eyeballs were in correspondence when I examined the patient.

In tumours of the brain, pain in the head is generally the first and most striking symptom, and next to it is affection of the sight. This affection of the sight has been found equally in cases in which the tumour had its seat in the cerebral hemispheres, and in cases in which the tumour was at the base of the brain. In tumours and other morbid states of the cerebellum or its membranes, affection of the sight has been rarely observed. Along with affection of the sight, transitory strabismus, slight paralysis of the face, or of one or other extremity, and transitory spasms in these parts, often simultaneously occur. Less frequently than disturbance of the motor power does disturbance of the sensibility manifest itself thus early. When disturbance of sensibility does occur, pains in one or other side of the face, or in one or other extremity, and the sensations of formication, " pins and needles," and numbness in different parts, are the forms in which it presents itself.*

* Dr. N. Friedreich, Beitraege zur Lehre von den Geschwuelsten innerhalb der Schaedelhoele. Wuerzburg, 1853, pp. 74, 75.

CHAPTER IV.

OF THE FAILURE OF HEARING FROM INJURY OF THE HEAD.

FAILURE of hearing from concussion of the brain is altogether similar in its pathology to amaurotic failure of sight from the same cause.

Fracture of the base of the cranium sometimes runs through the petrous portion of the temporal bone, and thereby breaks up the structure of the internal ear.

In respect to the degree of impairment of the function of the ear, it is to be observed that the hearing may be very much deteriorated from the natural standard without the patient being aware that he is deaf at all. On the other hand, in respect to objective symptoms, it is to be observed that notwithstanding deafness exists in a great degree, we may not be able to detect any corresponding morbid alteration of structure in the ears.

These remarks are equally applicable to cases of failure of hearing from injury of the spine.

PART THIRD.

DIAGNOSIS, PROGNOSIS, AND TREATMENT IN
CASES OF AMAUROTIC FAILURE OF SIGHT
FROM RAILWAY AND OTHER INJURIES OF
THE SPINE AND HEAD.

CHAPTER I.

PAIN in the head, confusion of mind, loss of memory, inability to fix the attention on anything, dulness of hearing, with *tinnitus aurium*, loss of muscular power of the limbs, with "pins and needles" sensations in the fingers and toes, which so commonly present themselves in the class of injuries before us, are symptoms indicating a profound disturbance of the brain as well as the spinal cord.

The tenderness and rigidity of the spine, to quote Mr. Erichsen again, the pain on pressing upon or on moving it in any direction, and the absence of any distinct lesion about the head, will sufficiently show that a case is one of spinal concussion. The two conditions of cerebral and spinal concussion, however, often co-exist primarily. The shock that jars injuriously one portion of the nervous centre, very commonly produces a corresponding effect on the whole of it—on brain as well as on cord.

The amaurosis arising from concussion of the spinal cord and that arising from concussion of the brain, we have seen reason to conclude, have for their common proximate cause a morbid state of the optic nervous apparatus.   But in the case of amaurosis from concussion of the spinal cord, the affection of the optic nervous apparatus has arisen by indirect transmission of morbid action from the original seat of the injury; whereas in the case of amaurosis from concussion of the brain, the optic nervous apparatus has suffered by being itself directly implicated.

CHAPTER II.

IN regard to this, there is no room for hesitation or doubt when a case has gone on to well-marked deterioration of structure visible at the fundus of the eye. But in those cases in which there is nothing more than some degree of vascular congestion of the optic disc and retina visible, and in which the symptoms are almost wholly subjective, viz., inability to exert the sight for any length of time, luminous and other spectra before the eyes, actual failure of sight, either transient or continuous, pain in the eyes or head, &c., there may be room for doubt, especially on the part of those not familiar with the varied and often apparently contradictory character of the symptoms which belong to amaurotic affections, or who have not directed their attention to the intricacies of the dependence of the healthy circulation in the optic nervous apparatus on the integrity of the cervico-dorsal portion of the spinal cord.

The patient, we have seen, may still be able to make out even the smallest letters, though he cannot con-

tinue to read longer than a few minutes. In such a
case, if the patient be a person below middle age, it
will be necessary to examine the eyes in order to determine whether they are hypermetropic or not. For if
so, the inability to continue to read might be owing
merely to an insufficiency of the adjusting power, as in
common asthenopia.

Perhaps the patient may only be able to make out
large letters. In such a case, if the patient be a person
past middle age, we must examine the eyes with a
view to ascertain whether or not the defect of sight be
owing merely to failure of the adjusting power in naturally emmetropic eyes.

If the inability to continue to read be owing merely
to common asthenopia, or the inability to make out
any but large letters be owing merely to presbyopia in
emmetropic eyes, convex glasses of the requisite focal
length will correct the defect. We have seen, however,
that in the cases under consideration, even should
common asthenopia from a hypermetropic conformation
of the eyes, or presbyopia in emmetropic eyes, exist,
and be corrected by the help of convex glasses, the
patient may still be unable to exert his sight on
account of the irritable congestion of his eyes and the
impaired energy of the optic nervous apparatus.

Perhaps again, it is with one eye only that the
patient can see to make out small letters; with the
other being able to distinguish large letters only. In
such a case it is necessary to remember that the fundus
of the better eye may present, under the ophthalmo-

scope, a greater degree of congestion than the eye of which the sight is so imperfect. This would show that the real cause of the imperfection is more deeply seated than the eyeball.

In the case of concussion of the eyeball above related at page 222, the disc of the uninjured eye was more congested than that which received the blow, and yet though both were affected with irritable asthenopia, the distinctness of sight was not impaired in either eye. Again, at page 220, Dr. Allbutt is quoted as having found in cases of slight tubercular meningitis, congestion of the optic disc and retinal vessels, though no defect of sight had been complained of; and at page 168, it is remarked that a patient affected with venous congestion of the optic disc may still have a considerable amount of vision. On the other hand, it is stated at page 167, that there may be failure of sight with little or no appearance of anything abnormal under the ophthalmoscope. Lastly, it is to be re-membered, that in cases of failure of sight, accompanied by well-marked atrophy of the optic disc, this degenera-tion may not, as observed at page 38, be the anatomical character of the morbid state on which the amaurotic symptoms actually depended at first, but rather the effect of that morbid state.

CHAPTER III.

PROGNOSIS IN CASES OF AMAUROTIC FAILURE OF
SIGHT FROM INJURY OF THE SPINE.

SOMETIMES, as before remarked, the affection of the
sight, though not less serious in degree, has been comparatively secondary in importance owing to the severity of the other injuries entailed on the unfortunate
victim of the accident. The question in such a case is
not, "Is the patient likely to recover the use of his eye-
sight ?" but rather, "Is he likely to be any better than
a broken-down invalid, who indeed may not immediately die, but who will never be the man he was
before ?"

As to the affection of the sight, even when the
general symptoms are not of so serious a character, the
question is often less the probability of a recovery,
than merely as to whether further deterioration of sight
can be prevented. From what I have said of the nature
of the morbid process going on in the optic nervous

apparatus, it will be seen that our prognosis, even thus conditioned, must be as unfavourable at least as in idiopathic cases of incipient amaurosis. And for this reason :—In our cases, the morbid process in the optic nervous apparatus is, as we have seen, a reflex from the disturbance excited in the spinal cord, so that even if it were capable itself of cure, the original cause—the affection of the spine—subsisting would prevent it.

The mention of spinal affection as in a degree the persisting as well as the original cause of the ocular affection, brings me in the last instance to the prognosis of the spinal affection. On this, however, I will not dwell, but shall make from Mr. Erichsen's able work, from which I have already quoted so much, the following extract, in which he has touched the question with a needle's point—*rem acu tetigit.*

" When a person has received a concussion of the spine from a jar or shake of the body, without any direct blow on the back, or perhaps on any other part of the body, and the symptoms have gradually and progressively developed themselves, the prognosis will always be very unfavourable. And for this reason : that as the injury is not sufficient of itself to produce a direct and immediate lesion of the cord, any symptoms that develope themselves must be the result of structural changes taking place in it as the consequence of its inflammation ; and these secondary structural changes being incurable, must, to a greater or less degree, but permanently, injuriously influence its action.

" The occurrence of a lengthened interval,—a period of

several weeks for instance,—between the infliction of the injury and the development of the spinal symptoms, is peculiarly unfavourable, as it indicates that a slow and progressive structural change has been taking place in the cord and its membranes, dependent upon pathological changes of a deep-seated and permanently incurable character."

PROGNOSIS IN CASES OF AMAUROTIC FAILURE OF SIGHT FROM INJURY OF THE HEAD.

Supposing the brain has not suffered material organic injury, the amaurotic symptoms occasioned by concussion of it will be found to yield to treatment more readily than those which supervene in cases of spinal concussion. A stun or slight blow on the head, however, has been known to be followed years after by loss of sight. Such were cases of cerebral amaurosis —an effect, among other symptoms, of disease of the brain, which had slowly and insidiously supervened, and which eventually caused death.

After general recovery from injury of the head, there may still remain impaired sight, immobility of the pupil, and paralysis of the muscles supplied by the third nerve.

CHAPTER IV.

TREATMENT OF AMAUROTIC FAILURE OF SIGHT, CAUSED BY RAILWAY AND OTHER INJURIES OF THE SPINE AND HEAD.

In approaching the subject of treatment, it is scarcely necessary to remark that attention must be directed, in an especial manner, to the spinal or cerebral affection caused by the injury. It will be a useful guide to us, therefore, to premise an inquiry into the therapeutical action of the principal remedies which have been found, by experience, efficacious in the treatment of corresponding idiopathic affections of the spinal cord and brain.

THERAPEUTICAL ACTION OF THE MEDICINES IN COMMON USE FOR AFFECTIONS OF THE SPINAL CORD AND BRAIN.

Belladonna.—In cases in which belladonna has been taken in poisonous doses, the following symptoms have been noted :—Small pulse ; dryness of the mouth and throat; paleness, succeeded by flushing of the face, distension of veins, and redness of the skin, like the rash of scarlet fever, with itching ; bluish red injection of the conjunctiva; coldness, followed by heat, with perspiration; mental disturbance. These symptoms indicate constriction of the small arteries of the surface of the body, with diminished afflux of blood,—the constriction of the arteries being owing to contraction of their circular muscular coat, excited by the belladonna. Thus, the small pulse speaks for itself :—the dryness of the mouth and throat, which is commonly an early symptom, no doubt arises from the constriction of the arteries of the mucous membrane of the fauces, and consequent suppression of its secretion; —the paleness of the skin arises from the impeded afflux of blood through the constricted arteries ;—the flushing of the face, and rash over the body, which succeeds the paleness, is to be accounted for by an accumulation of red corpuscles in the capillaries and venous radicles, which gradually takes place in consequence of the opposition offered by the constriction of the arteries to the free onward flow of blood ;—the blue injection of the conjunctiva is due to the same cause—venous congestion—as just explained ;—the coldness is owing to the diminished afflux of arterial blood to the skin at first, and the subsequent heat to the general venous congestion, as is also the perspiration. To the diminished afflux of arterial blood to the brain and spinal cord, with the consequent venous congestion, dependent on the constriction of the arteries, the cerebral disturbance and the diminution or loss of muscular power are to be ascribed.

By microscopical observation and experiment, I have

shown that the action on the arteries of certain sub-
stances commonly called stimulants, is to produce a
contrary effect to that of belladonna, viz., dilatation of
the arteries with a brisker flow of blood.* With these
facts in view, let us now ask: What are the agents
which have been found to operate as antidotes to the
poisonous action of belladonna? The answer is, stimulants such as ammonia and brandy, as also opium, &c.,—
the very agents which, when duly applied, cause dilatation of the arteries and a freer circulation of blood.
It is to be remarked, however, that stimulants such as
those mentioned, though they may mitigate the symptoms, are not really *specific* antidotes to the poisonous
action of belladonna. •

Under the action of irritating applications to the
conjunctiva, the pupil becomes contracted for the
time, but such transitory action on the pupil, it is
above shown (p. 189), is not the *specific* counterpart of
that exerted by belladonna. The only known agent
which specifically excites contraction of the pupil, as
belladonna causes dilatation of it, is, as already stated,
Calabar bean. In like manner it has been rendered
most probable by the experiments of Dr. Fraser, of
Edinburgh, that Calabar bean is a really specific antidote to atropia; as atropia is to Calabar bean.

In proof of the antidotal power of morphia against
atropia, Mr. Benjamin Bell, of Edinburgh, communicated in 1858 to the Medico-Chirurgical Society of

---

* See my Lectures on Belladonna, in the *Medical Times and
Gazette* for January 10 and 24, 1857.

that city a paper* in which he adduced two cases of atropism in which the subcutaneous injection of a solution of morphia was followed by rapid and striking abatement of the alarming symptoms. Though there can thus be no doubt of the value of morphia as an antidote to belladonna (and most likely *vice versâ*), its power is probably not of a directly specific nature, any more than is Vinum opii, in its action on the pupil, to be viewed as directly and specifically antagonistic to belladonna.

Led by the knowledge of the action of belladonna in exciting contraction of blood-vessels,† Dr. Brown-Sequard employed that agent* in cases of paraplegia due to a simple congestion or a chronic inflammation of the spinal cord and its meninges, and obtained a greater success from it than he had dared to hope for. Whatever be the value of our experiments, he continues, on animals, as regards the mode of action of belladonna, it is now certain that it has really a great power in diminishing the amount of blood in the spinal cord and its membranes. It is very well known that many

---

* "The Therapeutic Relations of Opium and Belladonna to each other." By Benjamin Bell, F.R.C.S.E. In *Edinburgh Medical Journal*, vol. iv. p. 1, 1859.

† "Lectures on the Diagnosis and Treatment of the principal forms of Paralysis of the Lower Extremities," 1861, p. 79. I have above shown (p. 194) that it is the arteries only, as I first discovered, which directly become constricted under the action of belladonna, and that Dr. Brown-Sequard has misinterpreted the phenomena, when he speaks as if belladonna excited contraction of all the vessels—capillaries and veins, as well as arteries.

French physicians, especially Bretonneau, Payan, Barbier and Trousseau, have for many years employed, with success, belladonna and ergot of rye in cases of paraplegia. In short, according to Dr. Brown-Sequard, belladonna is one of the most powerful and reliable remedies in cases of spinal congestion, meningitis, or myelitis in which there are symptoms of irritation; while on the contrary it is calculated only to increase the paralysis in cases of white or non-inflammatory softening of the spinal cord, which is characterized by the absence of symptoms of irritation.

From my experience of belladonna, it results that in cases of internal venous congestion like that which is of such frequent occurrence in persons affected with glaucoma, belladonna would operate prejudicially.

As Dr. Brown-Sequard points out, the symptoms of irritation which accompany spinal congestion and inflammation, constitute an important indication for the safe employment of belladonna. Such symptoms of irritation, it will be observed, correspond to those in scrofulous ophthalmia on which I have stated (at p. 136) belladonna operates so beneficially.

Belladonna may be given in the form of powdered leaves (gr. i—ii), or of extract (gr. $\frac{1}{4}$—$\frac{1}{3}$), or of tincture ($\mathfrak{m}$ x—xx) twice a day. Dr. Brown-Sequard also recommends belladonna to be applied over the spine in the form of a plaister six or seven inches long by four broad.

Ergot of Rye. According to Dr. Brown-Sequard the ergot of rye excites contractions of the walls of

R

blood-vessels,* and thereby causes constriction of their
calibre as belladonna does. The use of the ergot of
rye is indicated, Dr. Brown-Sequard says, in the same
cases as that of belladonna, namely, in spinal congestion and inflammation with symptoms of irritation.
On the contrary, in non-inflammatory softening of the
spinal cord the ergot is, as well as belladonna, contra-
indicated.

· The ergot is given in the form of powder, in doses of
from three to six grains twice a day.

Stramonium, hyoscyamus, and *Indian hemp* act in
a manner more or less like belladonna and ergot.
Against insomnia in cases of spinal congestion or inflammation with irritation, hyoscyamus or Indian hemp
ought to be employed instead of opium, the use of
which is dangerous under the circumstances, as it
increases the congestion of the spinal cord.

Calabar bean.—The action of this most remarkable
and interesting therapeutic agent in exciting the contraction of the sphincter pupillæ and the muscular coat
of veins, has been above described at pp. 189, 196, and
compared with the opposite action of belladonna, that,
namely, of exciting the contraction of the dilator pupillæ and the muscular coat of arteries.

This antagonism in the action of Calabar bean and
belladonna was calculated to suggest the inquiry as to

---

* I presume *arteries.* The likelihood that such would be the case
occurred to me many years ago, from a consideration of the action of
the ergot in exciting contractions of the muscular walls of the uterus,
but I never put it to the test, as I always found belladonna sufficient
for my purpose.

whether these two poisons might not prove specific antidotes of each other. The question as regards atropia *versus* physostigma, there is reason to believe, has been solved by Dr. Fraser, who, in a note communicated to the Royal Society of Edinburgh on the 31st of May last, relates some experiments in which rabbits and dogs survived the administration of *fatal doses* of Calabar bean, when certain doses of atropia were administered at the same time. One dog received eight grains of sulphate of atropia and six grains of dry extract of physostigma, by subcutaneous injection, and the animal recovered, after suffering from by no means severe symptoms, in less than four hours. Some weeks afterwards this dog received *three grains* of the same preparation of physostigma, also by subcutaneous injection, and *he died in seventeen minutes afterwards.*

The question as regards physostigma *versus* atropia appears to have been at the same time solved by Dr. Fraser's first experiment, for I presume that eight grains of the sulphate of atropia would have poisoned the dog, had it not been for the dose of physostigma at the same time exhibited. In short, the antidotal action of the atropia and physostigma would in this case appear to have been reciprocal.

The antagonistic action between belladonna and Calabar bean leads us to anticipate that the latter is likely to prove of use in those cases of intraspinal disease in which belladonna has been found by experience injurious. Venous congestion of the spinal

cord is the morbid condition in which Calabar bean
is likely to prove beneficial.*

In a case of tetanus arising from one of the toes
being accidentally cut off with a hatchet, my friend,
Mr. Wood, of Shrewsbury, divided and excised a portion
of the posterior tibial nerve. This had for effect an
abatement of the tetanic symptoms for a period of
forty-eight hours. But as they came on again to an
alarming degree, Mr. Wood had recourse to subcutaneous injection of one-third of a grain of the extract of
Calabar bean in solution every three hours. Under
this treatment the symptoms subsided, and the patient
recovered. The man was half-witted, and therefore
exercised no control over the spasms. He could not
open his mouth, nor protrude his tongue, nor could he
swallow. The muscles of the abdomen were rigid, and
hard as a board. The back was arched, and constantly drawn back by the spasms, so that opisthotonos
presented itself in the most marked form. The legs
and arms were cramped, and the feet drawn inwards.
The first effect of the injection of the Calabar bean was
relaxation of the arched back, then restoration of the
power of swallowing, and partially that of protruding
the tongue through the half-opened jaws. Relaxation
of the facial muscles also took place. There was
marked contraction of the pupil; but as the man had
had a dose of opium, this narcotic may have promoted the myosis. The sphincter ani was relaxed, and

---

* Calabar bean might be tried as a remedy in ague, and even in
cholera.

the motions were passed involuntarily. After the man got over the tetanic spasms, and was convalescent, he was attacked with thecal inflammation and abscess of the ring-finger of the left hand, and as this was getting well, the corresponding finger of the right hand became the seat of inflammation and abscess also.

In the case of paralysis of the third nerve with the pupil in a middle state, mentioned above at p. 177, the Calabar bean applied to the eye, at first merely for the purpose of testing its action on the sphincter pupillæ, proved in the end the means of cure. The following is a report of the case :—

The patient, E. W., widow, aged 41, a charwoman, was admitted into University College Hospital about the middle of May, 1869, as an out-patient, and after-wards, on the 31st of May, as an in-patient. She laboured under complete paralysis of the third nerve, left side, of rheumatic character.

Had rheumatic fever two years before, and was laid up therewith for three or four months. About the middle of April of the present year she was exposed to a draught of air, and almost immediately after noticed that there was something the matter with the left eye, as she could not raise the upper eyelid.

On examination of the affected eye, I found that the patient could not raise the upper eyelid, nor turn the eyeball towards the nose, nor move it upwards or down-wards. Besides this, the pupil was in the middle state of dilatation, and incapable of contracting under the stimulus of light.

When the eyelid was kept raised objects were seen double.

The patient complained of pain over the eyebrow, in the temple, and in the side of the head.

When the case first came under my notice, I dropped a solution of the sulphate of atropia into the left eye for the purpose of testing the action of this agent on the *dilator pupillæ.* The ensuing dilatation had passed off, and the pupil had recovered its previous middle state of dilatation before the patient presented herself again.

She was now taken into the Hospital, and put on good diet, as it appeared that she had been living very poorly of late. At the same time, she was ordered a pill containing calomel gr. j, extract of colchicum gr. ss, and Dover's powder grs. ijss., night and morning.

A solution of the extract of Calabar bean also was for the first time dropped into the affected (left) eye for the purpose of testing the action of that agent on the paralysed sphincter pupillæ. In making the application the idea glimmered through my mind that the Calabar bean might possibly exert a beneficial action on the paralysed nerve.

The effect of the application of the Calabar bean, on the 31st of May just mentioned, was to contract the pupil to the size of a pin's head in the usual time. On the 2nd of June, the action of the extract of Calabar bean on the pupil had passed off, but it was found, to my great satisfaction, that the patient had acquired some power over the upper eyelid, so far as to raise it a little by a strong effort. Seeing this, I repeated the application of the Calabar bean and discontinued the other treatment.

The quantity of Calabar bean used at a time was about one-fourth of a grain of the extract (an extract for which I am indebted to Dr. Fraser, of Edinburgh,) reduced to the state of an emulsion by the addition of a drop or two of water. This was freely poured into the eye.

On the 4th of June the patient could raise the upper eyelid a little more. The Calabar bean drops were accordingly repeated and continued thereafter three times a week. Under this treatment, I found at every visit to the Hospital, that the power over the levator palpebræ had increased. On the 14th of June, the report states that the patient could raise the upper eyelid sufficiently for the purposes of vision, and that she could move the eyeball both inwards and upwards, but only a little downwards. There was still double vision.

By the 21st of June, the patient had acquired nearly complete command over all the muscles of the eyeball, except the inferior rectus, and could raise the upper eyelid as well as that of the right side, and there was but slight diplopia.

On Monday, June 28, it was reported that Mrs. W. could raise the upper eyelid well, could turn the eyeball towards the nose and upwards well, but could not quite fully turn the eyeball downwards. On looking in this direction, therefore, she saw double.

Wednesday, June, 30. All the movements of the eye perfect, and no double vision, except in a slight degree, on looking downwards.

In regard to the pupil, it is to be remarked that it became contracted to the size of a pin's head at the usual interval after each application of the Calabar bean, but again returned to the middle state, when the influence of that agent had, as usual, passed off. The sphincter pupillæ has not, like the levator palpebræ and muscles of the eyeball supplied by the third nerve, as yet regained its power permanently, so as to contract under the influence of the action of the light on the retina. It is to be noted that the pupil of the right eye is in a nearly similar state, being but little moveable under the stimulus of light.

The adjusting power for the vision of near objects is impaired in both eyes, so that the patient requires the help of convex glasses, sixteen inches focus, to read with. The sight has otherwise remained good.

It is above observed, at page 185, that the sympathetic nerve, at the same time that it governs the consensual dilatations of the two pupils, appears to have some influence on the action of the external recti muscles, which is consensual therewith. A consideration of this led me to try the effect of dropping atropia into the eye in a case of paralysis of the external rectus, as soon as I had found, in the case just related, that the paralysis of the levator palpebræ and other muscles supplied by the third nerve yielded so decidedly to the influence of Calabar bean locally applied.

The case of paralysis of the sixth nerve referred to, in which I tried the local application of atropia, was not so favourable for the experiment as the case of paralysis of the oculo-motor nerve treated by the Calabar bean, because the paralysis in the former case was of intracranial origin, whereas in the latter case it was of a rheumatic character.

Mr. T. O. consulted me on account of his seeing objects double. He told me that he had been under treatment for severe pains in his head, and that latterly his sight had become confused. On examination, I found that the external rectus muscle of the left eye was paralysed. As long as he looked to the right, or straight forward, he saw objects single ; but as soon as he tried to look to the left, he saw double.

Under the use of sulphate of iron (gr. ij) and extract of nux vomica (gr. $\frac{1}{4}$) twice a day, the patient improved greatly in general condition, and recovered a considerable amount of power over the external rectus. The inability to turn the left eyeball towards the temple was, however, still very considerable, when, omitting all other treatment, I first confined myself to the local application of a solution of atropia (gr. iv—ʒi).

The first application made was for the purpose of dilating the pupil for an examination of the eye under the ophthalmoscope. Finding, a few days after this, an appreciable increase of the power to turn the eye outwards, and being assured by the patient that the two images of an object looked at were not so far apart as before, I continued the use of the atropia. And now, after a month's treatment with this agent, the last report is that there is scarcely any diplopia remaining, except when an attempt is made to look very much to the left.

Mercury.—As both mercury and iodide of potassium are medicines of the greatest value in the treatment of syphilis, they are not unfrequently considered as similar in their mode of operation, and that therefore the one may be substituted for the other indifferently. It must, however, be remembered, that it is in different stages of the disease that mercury and iodide of potassium are beneficial. In acute plastic inflammations, such as iritis for example, when the system is brought under the influence of mercury, the inflammation is in general observed to abate, and as this abatement goes on, the effused lymph becomes absorbed. Calomel combined with opium is the well-known form in which mercury is employed for the purpose mentioned. In chronic inflammations the bichloride of mercury in combination or not with bark, is the form indicated.

The action of mercury is commonly viewed as directly sorbefacient, but it rather appears to be sorbefacient because it subdues the inflammatory congestion on which the exudation of lymph depends, and the subsidence of which constitutes the proper condition for absorption.

Dr. Brown-Sequard recommends the use of mercury only in syphilitic paraplegia, and warns against its use in white softening of the spinal cord. In the chronic spinal meningitis and myelitis in our cases of spinal concussion, mercury in the form of the bichloride, in combination with bark, we shall see is of the greatest value.

Iodide of potassium has been used as a remedy in iritis and similar plastic inflammations, mainly under the idea that the absorption of lymph is promoted by it. The contrary of this, however, I have found to be the case. The effect of the iodide is to consolidate and fix the lymph in the pupil. It is in promoting the absorption of effused serous fluids that the sorbefacient power of the iodide of potassium consists, as has been well pointed out by Dr. Brown-Sequard; whereas it is lymph, the absorption of which is so much promoted by mercury, as above explained.

The iodide of potassium is, according to Dr. Brown-Sequard, the only known remedy that may be employed without danger in the various forms of paraplegia. It is especially useful in cases of white softening of the spinal cord due to fatty degeneration of the blood-vessels of that organ. It has, more than mercury, the power of producing the absorption of effused fluids in the vertebral canal. Indeed it is one of the most powerful agents in promoting the absorption of fluid effused in the cranio-vertebral cavity, either external to, or within the substance of the nervous centres. In cases of syphilitic paraplegia, its curative influence is sometimes very rapid.

The iodide is given in doses of six grains twice a-day. If there are signs of considerable effusion in the spinal canal in chronic meningitis, Dr. Brown-Sequard advises diuretics to be used in conjunction with it.

In combination with iron the iodide of potassium is

of great value in chronic subacute spinal inflammation as well as in non-inflammatory softening.

Bromide of potassium.—The bromide appears to act in a manner similar to the iodide of potassium, but does not like it affect the mucous membrane of the nostrils by inducing the symptoms of coryza. It has a soothing influence on the nervous system. Like the iodide of potassium the bromide is advantageously combined with iron.

Strychnine and nux vomica.—Dr. Brown-Sequard considers strychnine to be the only medicine that really deserves confidence as a means of increasing nutritive action in the spinal cord, and thereby increasing its vital energy.

Strychnine is not an excitant of the spinal cord. It is only by increasing the reflex faculty of the cord that strychnine seems to cause convulsions. So long as an animal under the influence of strychnine is left at rest, there are no convulsions, but the slightest touch immediately excites them.

The indications for the employment or non-employment of strychnine are:—

1st. Strychnine ought to be employed only in those cases of paraplegia in which there is no sign of irritation, or of increase of the vital properties of the spinal cord, such as the cases of reflex paraplegia, and of white softening of the spinal cord.

2nd. Strychnine ought to be avoided as a most dangerous poison in those cases of paraplegia in which there are signs of congestion or inflammation of the spinal cord, or its meninges. In such cases strychnine can only increase the cause of the paralysis.

The dose of strychnine may be $\frac{1}{40}$th—$\frac{1}{30}$th—$\frac{1}{20}$th of a grain daily; of nux vomica, gr. $\frac{1}{4}$—$\frac{1}{2}$. The use of strychnine should be suspended on the occurrence of spasms.

Iron may be advantageously given along with strychnine or nux vomica.

Iron.—Chalybeate preparations are of very extensive and beneficial application in our class of cases when, as generally happens sooner or later, the system has fallen into a low anæmic state. The iron is given in combination with bromide of potassium, or iodide of potassium, or nux vomica, as just indicated.

Dr. Brown-Sequard thus summarises the application of remedies in the different forms of paraplegia :—1st. In cases of paralysis of the lower limbs, with symptoms of irritation of the motor, sensitive, and vaso-motor nerve fibres of the spinal cord, or of the roots of its nerves, the appropriate remedies are the following :— Belladonna, ergot of rye, hyoscyamus, stramonium, Indian hemp, dry cupping, counter-irritation, the hot douche, and also sometimes the iodide of potassium, ammonia, sulphate of quinine, iron, or cod-liver oil.

2nd. In cases of paraplegia without symptoms of irritation of the spinal cord or of the roots of its nerves, the appropriate remedies are :—Strychnine, sulphur, the cold douche or shower bath, and also the iodide of potassium, and frequently ammonia, quinine, and iron.

It is of the utmost importance in the treatment of paraplegia not to make use of strychnine, belladonna, mercury, &c., before ascertaining positively whether there are or are not symptoms of irritation of the spinal cord.

TREATMENT OF THE PRIMARY EFFECTS OF SPINAL AND CEREBRAL CONCUSSION.

In the early stage of a case the first thing to be attended to is such a position of the patient as to ensure complete and absolute rest to the injured part of the spine, and to obviate every uncalled for movement of the body. Not less necessary is it that the mind should have rest in such cases on account of the tendency of the brain to become secondarily implicated even when it has not participated in the concussion which has affected the spine. In cases of injury primarily implicating the brain, of course every disturbing impression from without ought to be carefully guarded against.

The position of the patient should be as much as possible on the face or side to counteract determination of blood by gravitation to the spine and to facilitate the application of local remedies, such as plaisters, blisters, &c.

On the occurrence of reaction from the first effects of a shock, the propriety of abstracting blood by cupping or by leeches to the spine will require to be considered. If the condition of the patient be such as to forbid the abstraction of blood, dry cupping along the back on either side of the spine will be useful.

When the secondary effects of concussion of the spine begin to develop themselves, rest, Mr. Erichsen well observes, must, as in the early stages, be persevered in ; but in addition to this, counter-irritation may now be advantageously employed.

" With regard to internal treatment," Mr. Erichsen continues, "I know no remedy in the early period of the secondary stage when subacute meningitis is beginning to develope itself, that exercises so marked or beneficial an influence as the bichloride of mercury in tincture of quinine or of bark. At a more advanced period, and in some constitutions in which mercury is not well borne, the Iodide or the Bromide of Potassium in full doses will be found highly beneficial."

In the beginning of the treatment of chronic myelitis, Dr. Brown-Sequard employs the ergot of rye alone internally, and belladonna externally in the form of a plaister to the spine over the painful spot. The dose of the ergot, when the powder is used, is at first three grains twice a-day, gradually increased to six grains. The belladonna plaister must be a very large one, four inches wide and six or seven inches long. If there be no amelioration in a few weeks, the extract of bella-donna is to be given in doses of a quarter or third of a grain twice a-day. When, after six or eight weeks' treatment with ergot of rye and belladonna, improve-ment has not taken place, Dr. Brown-Sequard then gives the iodide of potassium in doses of five or six

grains twice a-day in addition to the preceding remedies; but when he has reason to suspect that there is
some degree of meningitis together with myelitis, he
begins the treatment at once with the iodide of potassium in addition to the ergot and belladonna.

When all signs of inflammatory action have subsided,
and when the symptoms have resolved themselves into
those of paralysis, whether of sensation or of motion,
but more especially in those cases in which there is a
loss of motor power, with a generally debilitated and
cachectic state, the preparations of nux vomica, of
strychnine, and of iron, may be advantageously employed. But the use of these remedies, more especially
of strychnine, must be particularly avoided in all those
cases in which inflammatory action is still existing, or
during that period of any given case in which there are
evidences of this condition.

Similar treatment is applicable in cases in which the
head has been the chief seat of concussion.

TREATMENT OF THE FAILURE OF SIGHT, SUPER-
VENING IN CASES OF SPINAL AND CEREBRAL
CONCUSSION.

In the treatment of the eye symptoms which
supervene in cases of spinal and cerebral concussion,
I have found the iodide of mercury in minute doses
(gr. $\frac{1}{20}$th—$\frac{1}{10}$th twice a day) beneficial, and after it,
the iodide or bromide of potassium with iron, followed
up by iron and nux vomica.

In the chapter on prognosis it has been remarked

that the amaurotic symptoms in cerebral cases have been found more amenable to treatment than those supervening in cases of spinal concussion.

In illustration of the course and treatment of spinal and cerebral concussion, with supervening affection of the sight, the following case is given in detail :—

CASE.

Mr. T. G., aged 36, in a railway accident on the 16th of November, 1868, received a blow both on the back of the head and on the spine in the lumbar region, and suffered a general shock of the nervous system. He was in consequence confined to bed for nearly three months.

When he first consulted me on the 4th of February, 1869, his right leg was very weak and lame, and he dragged the foot in such a manner as to scrape the ground with his toe, so that the sole of his shoe was quickly worn away at the point. His right leg had been four degrees below the temperature of the other, and he complained of tingling and pricking sensations in his fingers and toes. His mind and memory were somewhat confused, and the hearing of his left ear dull.

In regard to his sight, the patient stated that about six weeks after the accident the right eye became affected, and shortly after that the left also. The sight of the left eye was now worse than that of the right.

s

On examination, I found the upper eyelid of the left eye in a state of incomplete ptosis ; both eyes dull, and watery-looking ; the rectal veins large and numerous ; the pupil of the left eye smaller than that of the other, but both obedient to the light, though not active in their movements. The impression of light appeared to be painful, especially to the left eye.

The state of the sight was as follows :—When a finger was held up before him to the left, Mr. T. G. said that he saw it double, the two images being side by side, and about half an inch apart. When the finger was moved to the right he did not see it double. This diplopia appeared to be owing to impaired power of the left external rectus, but in so slight a degree was the non-correspondence of the visual axis of the two eyes, that any deviation from their normal direction on looking to the left was scarcely perceptible.

Formerly the sight was good for any distance, and could be exerted for any length of time. Now he cannot look at anything for more than a few minutes; with both eyes open cannot see anything so distinctly as when one eye is shut ; when the left eye is shut, can see to read with the right a few words at a time ; when the right eye is shut, cannot see to read with the left. In reading with the right eye, the left being closed, he holds the book near, and appears to strain the eye all about. Convex glasses of from twenty-four to eighteen inches focus assist the sight a little.

Has not seen motes or luminous appearances before his eyes.

Under the ophthalmoscope, the pupil of the right eye remained rather wider than that of the left. The right eye bore the examination pretty well, but the left suffered greatly from pain, both at the time and after the examination. In the right eye the retina was seen to be much congested, the veins gorged, and the disc the seat of capillary injection. In consequence of the sensitiveness of the left eye to the light, I was able to obtain no more than a glimpse of its fundus. This glimpse, however, was sufficient to satisfy me that the optic disc and adjacent part of the retina were in a state similar to that in which I found the corresponding parts in the right eye, though the congestion was in a greater degree.

The drooping of the upper eyelid, and the double vision on looking to the left side, with the paralytic halt in the right leg in the case related, appear to be referable to some intracranial cause on the left side.

Mr. T. G. has been under the care of Mr. Erichsen and Dr. McOscar since the accident, and under their treatment has recovered so far as to be able to go about. The disturbance of the sight, however, continuing, I was called into consultation, and the patient has since continued under our joint care.

February 6, 1869. — Was much fatigued by the examination of the eyes on the 4th, and after returning home had a feeling as if he was about to retch, and saw stars of light before him. These appearances again presented themselves before him this morning, and in crossing the street his sight went from him for a minute or so. The eyes and sight are altogether worse to-day. The following prescription was agreed to :—

> ℞—Hydrargyri Bichloridi gr. j.
> Tincturæ Iodi ʒj.
> Potassii iodidi ʒjss.
> Aquæ ad ʒx.
> F. mistura.

Sig.—One tablespoonful twice a day in decoction of sarsaparilla for the next ten days.

February 16.—The patient thinks his eyes improved. The sight, however, has gone from him several times, and stars of light have been seen since last visit.

The intolerance of light and dimness of sight in the left eye continue ; and there is still the drooping of the left upper eyelid.

Double vision persists, with a strained and uneasy feeling when he turns the eyes to the left. On looking to the right, vision is single, and pretty distinct, but he cannot read without confusion coming on.

Has been suffering severe pain in the head as if it would burst, but the spine is less sensitive.

The medicine causes nausea. It was, notwithstanding, considered advisable to continue it.

February 25.—Thinks the eyes much better. The sight has not gone from him so much as it did before. Instead of a star has sometimes seen a dark spot.

Left eye less sensitive, and when the right is shut

can now see to read with it a little. The left upper eyelid does not droop so much.

Diplopy, on looking to the left, is now no more than an image and a shadow close to it. That is, if he turns the eyes slowly and circumspectly to the left. If he turns them suddenly to the left he then sees double in a more decided degree.

Can now see better with the two eyes open on looking to the right, but is still unable to read longer than a few minutes at a time.

Thinks the state of his mind and memory improved, and feels more cheerful in spirits, but still complains of the peculiar pain in the head.

Right leg very bad.

Under the ophthalmoscope I found the fundus of the right eye less congested generally. The veins were, however, still large and tortuous, and the disk reddish gray. The left eye, though it bore the examination better, was pained by the light. The congestion at the fundus of it was observed to be still considerable.

To continue the medicine.

March 9, 1869.—Not so well, either generally or in respect to the eyes. Two days ago the sight failed him entirely every two or three hours, for about half-an-hour at a time. Otherwise, the sight has been better. Does not now see double, except when he turns the eyes sharply to the left. Cannot continue to read any better. After reading a short time, the page has seemed to become black.

There has been no more photopsy.

Still drooping of the left upper eyelid.

Has a cold, which has upset him.

The left eyeball is tender to the touch ; the right not so sensitive.   Pupil of the left eye more susceptible to the light than that of the right eye.

Right leg still very bad.   Stiffness in the joints, and tingling in the hands.

Besides the pain in the back of the head, which still persists, complains of pricking pain in the crown of the head.

Pulse languid.   Bowels pretty regular.

To omit the iodide of mercury mixture, and to take the following instead :—

R—Potassii Bromidi ʒiij.
Vini Ferri ʒiii.
Aquæ ad ʒx.
F. mistura.

Sig.—One tablespoonful twice a day.

March 16.—Very much better to-day.   Drooping of the left upper eyelid nearly gone.   Pupils active, though they tend to dilate again after contracting.   No longer sees double on looking to the left, provided he turns the eyes cautiously in that direction.   Can see more distinctly, and read for a longer time.

Under the ophthalmoscope the fundus of the right eye—both disc and centre—appeared more natural. The left eye bore the light much better ; and it was seen that though the optic disc was still considerably congested on the nasal side, the appearance of the fundus was altogether improved.

When the right eye is closed can now see to read a little with the left. The photopsy and transitory failure of sight have not recurred.

Mind and memory more active and spirits better.

Sees across the street better with concaves No. 1 than with the unassisted eye, and reads more comfortably with convex glasses, 24 inches focus.

Continue the bromide and iron—a dose thrice a day.

April 6.—Still suffering from the effects of the cold.

Though the improvement as regards the drooping of the left upper eyelid and the double vision continues, the patient has fallen back in other respects. Neither his memory nor his spirits are so good.

Continue.

April 20.—Pains in the back of the head worse; lameness of the right leg worse. As to the eyes, they have again become weak, irritable, and congested-looking externally.

Under the ophthalmoscope the disc of the right eye appears so much congested as scarcely to be distinguishable from the adjacent retina. The left eye was too irritable to bear the ophthalmoscopic examination.

Drooping of the upper eyelid now scarcely to be noticed.

Double vision only on turning the eyes suddenly to the left, and that only for a short time, for by fixing the eyes on an object it goes off.

No transitory failure of sight nor photopsy.

Hearing still disturbed. Hears better with the two ears in use than with one singly.

Pressure over the spinous process of the 7th cervical vertebra causes uneasiness at the place, with a sudden discharge of pain as if shooting to the head.

Painted caustic tincture of iodine behind and below the mastoid processes for the sake of counter-irritation.

To omit the bromide of potassium and iron, and to take one-twelfth of a grain of the bichloride of mercury in decoction of sarsaparilla twice a day.

April 27.—Eyes better; the ptosis gone; the diplopy gone ; but still unable to read for any length of time.

Under the ophthalmoscope, the discs not so much injected as at last visit, though the congestion at the fundus is still great. The veins tortuous.

Hearing is still defective.

The general state has not improved. The pain in the head worse ; the mind confused ; the memory impaired ; the sleep bad ; the appetite indifferent.

Troubled with pins and needles sensations in the left hand.

Severe pains in both knees, but especially in the right popliteal region in the situation of the inner hamstring, thus aggravating his lameness.

Right leg numb, and twitches for some minutes when he gets out of bed.

Feels tired and languid. Pulse 96, and weak.

Omit the bichloride, and take one of the following pills twice a day:—

> ℞—Ferri Sulphatis gr. ij.
> Extracti Nucis Vomicæ gr. ⅓.
> Extracti Anthemidis gr. ij.
> F. pilula.

May 18.—Altogether much better both generally and in respect to the eyes. Is now able to read a paragraph or two of the newspaper. Left upper eyelid is now almost as fully elevated as the right. Objects seen single on looking to the left as well as on looking to the right. Both pupils equally active. Under the ophthalmoscope, the fundus now presents a more natural appearance generally, and the optic discs much less congested. The eyes, however, still look weak and watery.

Continue the pills of iron, with the extract of nux vomica increased to gr. ½ in each.

June 1.—Sight continues to improve. Under the ophthalmoscope congestion at the fundus is scarcely any longer appreciable.

Continue the pills.

June 15.—Last Sunday week put on summer clothing, and next day found himself numb in all his limbs, and has not been so well since.

Eyes appear much congested externally and are somewhat intolerant of light. Pupils active. Was seized with a fit of giddiness the other day so that he fell on his face.

Lameness of right leg continues.

Omit the pills and take nothing.

June 29.—Eyes continue to improve. The left is still irritable, and more congested externally than the right. Both pupils active. Reading soon tires the mind as well as the eyes.

Though the appetite is pretty good, does not feel much better as regards general condition.

To resume the pills of the sulphate of iron and nux vomica.

July 13.—Though the eyes still appear much congested externally, especially the left, the state of the fundus, as seen under the ophthalmoscope, has much improved.

Spirits are very much better, and, therefore, the symptoms generally are less complained of. Perspires freely.

Continue the pills.

July 27.—Very much better in every respect, the condition of the right leg still excepted. Just above the knee there is a feeling of tightness or cramp. In walking still scrapes the ground with the point of the right foot. The condition of the leg varies. Has had no return of vertigo, but still feels the pins and needles sensation in both hands, but especially in the left.

Spirits are good and the expression of the face is brighter.

The eyes are very much improved. No external

congestion remaining except the enlargement of the rectal veins. Pupils natural in size and action. Can read the newspapers in the morning without any great fatigue.

To go to Spa for a month or six weeks for change of air and to drink the waters.

APPENDIX:

COMPRISING AN ADDITIONAL CHAPTER ON
INFLAMMATION.

ADDITIONAL CHAPTER.

EXAMINATION OF SOME POINTS RELATING TO THE INFLAMMATORY PROCESS.

In the pathological exposition of the influence of the vaso-motor nerves on the circulation of the blood in the extreme vessels, the inflammatory process necessarily came under consideration, and a chapter in the text is accordingly devoted to it. It would have been out of place to have there entered more minutely into the subject. Certain points relating to the process were therefore, left untouched. The points referred to, however, on account of their general pathological importance, appearing to me still to require some notice, I here devote an additional chapter to their examination.

STATE OF THE BLOOD AND THE BLOOD-VESSELS IN THE FROG'S MESENTERY, EXCITED TO INFLAMMATION IN CONSEQUENCE OF BEING DRAGGED OUT ALONG WITH THE INTESTINE THROUGH AN OPENING INTO THE ABDOMINAL CAVITY.

My observations were made on frogs rendered insensible by being held for about a minute under water

of the temperature of 110° to 120° Fahr., as recom-
mended by the late Dr. Marshall Hall.

In the mesentery of the frog there is little of a
capillary network, and it may be said that many of the
last ramifications of the arteries, losing the muscularity
of their walls, and so far acquiring the character of
capillaries, open into the side of small veins.

In the walls of the intestine there is a close capillary
network to which the large arteries in the mesentery
run, and from which the corresponding large veins
return. This capillary network, and the arteries and
veins leading to and from it, I found to become gorged
with blood, but nothing further in detail could of
course be observed of the state of these vessels, in
consequence of the thickness and non-transparency of
the intestine.

In the transparent part of the mesentery intervening
between the great arterial and venous trunks—the part
on which alone microscopical observations can be made
—there is, as I have said, not much of a proper capillary network, but chiefly small arteries and small veins
opening into each other in the manner described.

At the place where the great arterial trunks reach
the intestine, and begin to sub-divide into the intestinal
branches, they give off also the arteries which ramify in
the transparent parts of the mesentery. The latter
arteries follow a *recurrent* direction away from the
intestine.

Into the venous trunks, in their course in the
mesentery from the intestine, the small veins corre-

sponding to, and receiving the blood from the small mesenteric arteries just mentioned, open here and there.

When the mesentery was first disposed under the microscope, the state in which I found the circulation in the blood-vessels of its transparent parts (the arrangement of which has now been described) was as follows:—The small arteries were, for the most part, rather dilated, and the flow of blood in them, as in arteries in a similar state of dilatation in the web, correspondingly rapid and free. Sometimes the arteries were seen in such a state, that without being constricted, they could not be said to be dilated. Sometimes, again, the arteries were found rather constricted, though never to closure, with a corresponding appearance of increase in the thickness of their muscular wall. This muscular wall, however, it is to be observed, was not so thick and strong as the muscular wall of the arteries of the same size in the web. It seemed to be of much the same thickness and strength as that of the correspondingly sized arteries in the lung which, as stated at page 91, I never actually saw become much constricted.

It is, as I have at p. 78 insisted, only when the blood flows into a dilated from a constricted part of an artery that retardation of the stream is observed. In the mesentery of the frog under the microscope I saw, for instance, in a large, thick-walled, arterial trunk, which was much constricted in one part of its course, and greatly dilated here and there in other parts, the blood become retarded in its flow on entering the dilated

parts; a state of matters in all essential respects similar
to what I have described and delineated in so much
detail in my often-quoted Essay in Guy's Hospital
Reports, pp. 8, 17, 21. The figure at p. 21 is, indeed,
almost an exact representation of the condition of the
artery in the mesentery I am referring to.*

In uniformly dilated arteries in the mesentery, on
the contrary, I found, when there was no impediment
to the circulation in the part from local pressure or
from disturbance of the heart's action, the flow of blood
was accelerated. Any retardation of the flow of blood
in uniformly dilated arteries in the mesentery of the
frog observed to take place, I found to be owing, in a
great measure, to the impediment occasioned by the
dragging and pressure to which the parts under the
microscope were liable to be subjected, for when I
relaxed and altered the arrangement, and so relieved
the dragging and pressure, the circulation again became
more free and rapid. The same thing I have over and
over again observed in the web. Another cause was
obstruction of the pulmonary circulation, and impeded
action of the heart.

It would be difficult if not impossible to estimate in
detail the effect on the circulation produced by opening

---

* I may here observe that the figure at p. 21 has been copied by
Professor Virchow in his Cellular Pathology, but is there erroneously
stated to be an artery constricted by the action of atropia. The
figure in question ought, in my essay, to have come in at the end
of the paragraph, but a crisis in the arrangement of the types
necessitated the introduction of the wood block immediately after
the mention of the constricting action of atropia.

the abdomen, and dragging out the intestine and mesentery, and disposing the latter under the microscope. Exposure to the air operates on the prolapsed intestine as an irritant, exciting in a great degree its peristaltic contractions. In my Essay in Guy's Hospital Reports, pp. 8, 9, I showed that under the operation of various influences on the web, the arteries become constricted or dilated—usually first constricted, and afterwards dilated,—and that, under certain circumstances, constriction either does not take place at all, or, when it does, it very rapidly gives place to great dilatation. Exposure to the air seems to operate on the arteries of the prolapsed intestine and mesentery in a similar manner.

An effect on the arteries of the prolapsed intestine and mesentery being dilatation, with or without previous transient constriction, the consequence of this dilatation is increased determination of blood to the parts,—both capillaries and veins becoming gorged with blood loaded with corpuscles. Such is *Congestion*. If now the flow of blood in the arteries be impeded by constriction or by pressure so that *vis à tergo* is diminished, *stagnation* takes place in consequence of the agglomeration of the red corpuscles in the capillaries and venous radicles, as explained in the text, p. 132.

Any tortuousness of the arteries more than natural observable, appeared to me to be owing merely to an uneven disposition of the mesenteric web under the microscope. The change from straightness to tortuousness and *vice versâ* from such a cause, is well seen in the pulmonary artery, according as the lung becomes collapsed or distended with air.

The dilated state in which the veins are found, it is to be observed, is a necessary consequence of the congestion in the capillaries of the intestine. If there had been free circulation through the capillaries of the intestine, distention of the veins with blood would have been owing to the preceding dilatation of the arteries in the manner described (at p. 80) in the rabbit's ear, when the artery dilates, and in the frog's web when the arteries are dilated from section of the ischiatic nerve. But the distention of the veins in the mesentery was owing, as I have said, to the congestion in the intestine from which the veins return. The distention of veins with blood under such circumstances takes place by slow accumulation, whereas, when the circulation is free in the capillaries, it takes place rapidly in proportion to the dilatation of the arteries.

The intestine and its mesentery being drawn out and disposed under the microscope as above described, it is the intestine which is the focus of the inflammation which ensues. The flow of blood in the transparent part of the mesentery, though it may not yet be obstructed, is very much influenced by the state of the stream of blood in the arteries going to, and in the veins returning from, the intestine. The large arteries going to the intestine being, as before stated, here and there constricted, and the returning veins very much gorged.

In the small veins in the clear part of the mesentery

colourless corpuscles are seen to accumulate in considerable quantities on the walls, in the same manner as seen in the veins of the web; and according to the retardation of the flow of blood in the vessels, so is the accumulation of colourless corpuscles increased. The stream from the less dilated arterio-capillary vessel, on entering directly, in the manner above described, the side of a wider vein, immediately becomes much slower, and numbers of colourless corpuscles are seen at once to subside on the walls of the vessel.

As in the vessels of the web, the colourless corpuscles appeared in the vessels of the mesentery round and plump; except, perhaps, a few here and there, which, adhering to the wall of the vessel by a point, were pressed by the force of the stream into a pearshape (p. 127-8).

In the small veins of the mesentery, colourless corpuscles are thus seen to accumulate on the sides of the walls in great quantities, in the same manner as we see them in the veins of the web. In the web the accumulation of colourless corpuscles may become dispersed, and the flow of blood resume its natural characters, if the animal be set free from the cramped position in which it was under the microscope, and thus allowed to recover itself. The same may be said of the stagnation of red corpuscles which is often observed in a capillary here and there. But it is quite otherwise in the case of the intestine and mesentery dragged out of the abdominal cavity and exposed to external influences from which, in the natural state, they are so

carefully guarded. Here the result is the establish-
ment of a condition of the blood and the blood-
vessels similar to that in the part adjacent to a
slough in the frog's web, as described in the following
extract from my Essay in Guy's Hospital Reports, pp.
51, 52 :—

" As to the state of the blood, and the blood-vessels
generally, of a web inflamed in the degree above
described : The arteries and veins are dilated. The
veins very much dilated, and their walls sometimes
no longer distinguishable. On the sides of such dilated
venous channels, numerous colourless corpuscles may
be seen accumulated whilst the blood is flowing freely
in the middle. In vessels in which, in the course of
the healing process, the flow of blood has become
re-established, numbers of blood-corpuscles, red as well
as colourless, continue to adhere to the walls.

" In a case in which the second web had sloughed
away to a small piece, stagnation of blood still existed
at the edge of that piece ; but higher up, a free flow of
blood had now superseded stagnation. Here the walls
of many of the vessels can no longer be distinguished,
and the blood appears to be flowing in mere passages,
hollowed out in the substance of the web. All along
the sides of the passages, numerous blood corpuscles
are adherent. The adherent corpuscles are most of
them *red*, but most of these red ones, again, are *round*.
In vessels in which stagnation had not continued so
long, colourless corpuscles were seen in greater number,
but not much greater. In the portion of web under

notice, spots of extravasated blood were seen here and there."

In a frog, 72 hours after the abdomen was first opened and the mesentery drawn out for microscopical examination, I found the blood still flowing in two of the arterial trunks and in their corresponding veins, though with little force in consequence of the weakened action of the heart. The arteries were very much constricted and lined with quantities of colourless corpuscles, which were seen either stagnant on the walls or rolling sluggishly along. Some were seen adherent by a point, and after being drawn into a pear-shape became detached and were carried along in the stream. The veins were very much dilated and gorged with slowly-moving blood, loaded with red corpuscles. In all the vessels—arteries, artery-capillaries, and veins— in the transparent part of the mesentery, the blood was stagnant. A somewhat similar state of vessels observed in the mesentery of the mouse, I have recorded to the following effect in my Ophthalmic Medicine and Surgery, p. 557:—A varicose state of the venous radicles I saw take place in the mesentery of the living mouse under the microscope, in consequence of the walls of the vessel, here and there, becoming much constricted, and, here and there, much dilated. This was towards the end of the observation. The corresponding arteries were constricted, except where their last ramifications opened into the capillaries. Here the blood flowed into them in a retrograde direction, and sluggishly, from anastomosing vessels. In the varicose venous radicles there was congestion of red corpuscles. This observation illustrates the relation between constriction of the arteries and congestion of the veins.

AMŒBA-LIKE CHANGES OF FORM WHICH THE COLOURLESS CORPUSCLES UNDERGO IMMEDIATELY AFTER THE BLOOD IS DRAWN.

Preparatory to considering the comportment of the colourless corpuscles, within the vessels of an inflamed part, the changes of form they undergo in blood which has been just drawn from the body, require to be taken into account.

A small drop of blood from a prick in the skin being received on a slide, spread out by covering it with a thin scale of glass, and disposed under an eighth of an inch objective, we may observe the colourless corpuscles, on narrowly watching them, in the act of changing shape. Fixing the attention on one, we shall see it, perhaps, looking as if it had burst and was pouring out its contents; but this is not the case. The change of form is owing to a contraction of the cell-wall around its contents and nucleus on one side and a protrusion of it in finger-like processes on the other. The processes, in this case, look like some extraneous flake of lymph adhering to a small cell. By-and-by, however, it is seen that this protruded part of the cell-wall shrinks in and comes to look more like a part of the cell.

As the fibrin is deposited by coagulation from the plasma, and serum only remains surrounding the corpuscle, it re-acquires more of its previous globular form. This appears to be owing to imbibition into its interior of some of the less dense serum, for by the addition of a little water to the drop of blood under examination, we see that the corpuscle becomes completely distended and round.

The colourless corpuscles do not all undergo such a remarkable change of shape. There is a species of colourless corpuscle, named by me the coarsely granular cell on account of the large size of the granules forming its contents, which, though it may be observed to undergo some alteration of shape, does not shoot out its cell-wall into processes in so remarkable a degree.

The corresponding corpuscles in the mammifera undergo similar changes immediately after the blood is drawn from the living animal. See my Papers on "The Blood Corpuscle considered in its different phases of development in the Animal Series," in the Philosophical Transactions, Part II., for 1846, from which I make the following additional extracts :—

Colourless Corpuscles of the Blood of the Skate.— These corpuscles are about $\frac{1}{1800}$th of an inch in diameter, and are composed of an agglomeration of granules surrounded by a cell membrane. In consequence of this structure, I designate them granule blood-cells.

On one side of this granule blood-cell, a clear spot may sometimes be seen, indicating the place of a nucleus. By the action of dilute acetic acid the granules are dissolved and the nucleus comes into view.

I have above spoken of the granule blood-cell as being of a roundish form : as the blood of the skate, subjected to examination, was obtained from the blood-vessels of the viscera already removed from the animal, I cannot say what form the granule blood-cell would present in blood drawn from the animal during life, and forthwith examined. But in some examples of the blood which I examined, and which was obtained at a time which could not have been long after the death of the animal, the granule-cells at first presented most re-markable changes of shape and other phenomena im-mediately to be described, which after a time ceased to be presented, the cells settling into a round form.

As the changes of shape and other phenomena to which I refer, resembled appearances presented by what we shall find to be the corresponding corpuscle in the blood of the frog, and of other animals, drawn during life, I am disposed to believe that they will also be found presented by the granule-cells of blood actually drawn from the living skate.

My attention was first attracted to the phenomena by observing a granule-cell with the granules apparently escaping from it as if burst. But the cell soon appear-ing again with all the granules collected together, I was led to watch, and soon perceived that the appearance of granules escaping as if from a burst cell, was owing to this :—The transparent and colourless cell-wall bulged out on one side, leaving the granules still agglomerated and holding together; but this was only for a short time, for soon single granules were seen to separate and burst out from the rest, and to enter the hitherto empty compartment produced by the bulging out of the cell-wall. The regular manner in which this sometimes took place was remarkable. I have actually seen the granules enter the compartment by one side, and circulate along the bulging cell-wall to the other side, until the whole compartment became filled with granules. This having occurred, the bulging began to subside, but was succeeded by a bulging of another part of the cell-wall, into which again a flow of granules took place, and so on all round the cell.

Colourless corpuscles of the blood of the frog.—The corpuscle in the blood of the frog, clearly homologous with the granule blood-cell of the skate in the coarsely granular stage, is roundish, and contains clear granules strongly refracting the light, like those of the corresponding cell in the blood of the skate, only not so large. These granules are usually collected in greater numbers to one side. The cell itself I have seen, on carefully watching it in blood drawn from the living animal, exhibiting changes of shape with movements of the granules such as have been above described in regard to the granule-cell of the skate.

There is another form of colourless corpuscle which I designate the granule blood-cell in the finely granular stage : That of frog's blood rapidly undergoes change of shape, and, when first seen, presents processes shooting out from it.

Blood-corpuscles of crustacea.—If very great ex-
pedition has been employed in receiving the blood as it
flows from the claw of a crab cut across on the plate
of glass, spreading it out and transferring it to the
microscope for examination, the granule-cells may be
seen to be of an elongated oval shape, the nucleated cells,
spindle-shaped. These shapes, however, are speedily
changed.

As regards the granule-cell, it tends to become
circular, but it is also soon observed that its cell-wall
bulges out here and there into round processes which
again subside, whilst another part of the cell-wall bulges
out in the same way. This change of shape, it will be
perceived, is similar to that which I particularly described in the case of the granule-cell of the blood of
the skate, only it is to be remarked that it is not accompanied by such a well-marked movement of the
contained granules. Besides bulging out into round
processes, the cell-wall may be seen to shoot out into
cilia-like processes also.

As regards the nucleated cell, it also first tends to
become circular, and shoots out its wall into processes
which are usually more cilia-like than in the case of
the granule-cell; and, being in all directions like radii,
the cell comes to present a stellate appearance. Sometimes a cell is seen to shoot out into processes in two
principal directions only, these processes again shooting
out into smaller, so that the cell acquires a caudate
form.

The description which has now been given of the

blood corpuscles of the crab is in all respects applicable to those of the lobster.*

The blood corpuscles of other invertebrate animals undergo similar changes of shape, but it is not necessary to particularise them here.

CHANGES OF FORM WHICH THE RED CORPUSCLES OF THE BLOOD UNDERGO.

After the account just given of the amœba-like changes of shape which the colourless corpuscles undergo, there remain to be noticed such changes in shape as the red corpuscles are observed to present. The red corpuscles are yielding and elastic, so that they readily change shape when slightly pressed upon, like partially-filled bladders, which, indeed, they are, and as readily regain their original form when they have escaped from the compressing agent. In consequence of this property the corpuscles glide along in their vessels, with great ease accommodating themselves to all obstacles and to each other.

At pages 92 and 93, a difference in the appearance of the red corpuscles within the vessels of the web and those of the lungs is pointed out. Within the vessels of the lungs they are seen to be less flat-looking and more flexible and to accommodate themselves more easily to any obstacle.

---

* The changes of form which the blood corpuscles of the lobster, like those of the crab, undergo after the blood is drawn were described and delineated by Hewson as accurately as his microscope appears to have enabled him to observe them.

In inflammatory or buffy blood, the red corpuscles are also more flexible.

In a mixture of blood and pus the red corpuscles are observed to yield in the most extraordinary manner, so as to accommodate themselves to obstacles. Thus, in order to pass through a narrow channel, they will be drawn into a mere filament, and yet, when free, immediately regain their original form. This capability of being moulded into various shapes appears to depend on their state of distention, and this again on the nature of the liquid in which they are suspended.*

COMPORTMENT OF THE COLOURLESS CORPUSCLES IN THE VESSELS OF AN INFLAMED PART, AND THEIR ALLEGED ESCAPE THROUGH THE WALLS OF VEINS AND CAPILLARIES.

In a paper on Inflammation and Suppuration,† Dr. J. Cohnheim enunciates a view of these processes very different from that so earnestly advocated by Professor Virchow, his colleague, and Principal at the Pathological Institute in Berlin. In opposition to the author of "Cellular Pathology," Dr. Cohnheim recognises the essential dependence of the inflammatory process on a disturbance of the circulation, and repudiates the hypothesis of "Proliferation" altogether.

The microscopical observations to which Dr. Cohn-

* See my Observations on Some Points in the Anatomy, Physiology, and Pathology of the Blood, in No. 28 of the *British and Foreign Medical Review*, 1842.

† Ueber Entzündung und Eiterung, in Virchow's Archiv für pathologische Anatomie und Physiologie, for September, 1867.

heim chiefly appeals in support of his views regarding inflammation and suppuration, were made on the mesentery of frogs rendered insensible and motionless by the action of Curare poison. But apparently not taking into account all the varied phenomena of the circulation in the extreme vessels in the natural state, the constrictions and dilatations of the arteries especially, and the influence thereof on the flow of blood in the arteries themselves and in the capillaries and veins, Dr. Cohnheim has not, in my opinion, correctly interpreted the states in which he saw the blood and the blood-vessels in the mesentery of the frog under the conditions under which his observations were made.

I agree with him, however, when he repudiates the hypothesis of " Proliferation," and calls in question the reality of any process of cell-multiplication such as that by division of pre-existing cells, at least as regards pus-corpuscles.

The peculiarity of Dr. Cohnheim's view of inflammation and suppuration is, that the colourless corpuscles of the blood escape bodily from the interior of the blood-vessels of the inflamed part through pores in their otherwise entire walls, and accumulating, constitute the corpuscles of the pus which presents itself.

The accumulation of colourless corpuscles which he saw in the veins of the mesentery Dr. Cohnheim describes as an essential preliminary step in the inflammatory process. But it does not, he says, continue long, for an unexpected phenomenon occurs; viz., on the outside of the wall of the veins small boss-like diverticula present themselves, and gradually become more prominent until they appear like colourless corpuscles protruding half-way through the wall. Protruding still more, this body comes to present itself as a pear-shaped corpuscle outside the vessel, but still fixed to it by a stalk. From the body of this corpuscle processes shoot out. The stalk being drawn out longer and longer, the corpuscles at last become detached from the outside of the wall of the vein. In the smaller veins and in the capillaries, Cohnheim says he has seen colourless corpuscles half-way out and half still within the vessels, and at last wholly outside the walls.

The red corpuscles continue to be carried along in the stream, and none are seen to have escaped along with the colourless ones.

I must confess that, though I have continued to watch the circulation of the blood in the transparent part of the frog's mesentery for a long time, I have never been able to satisfy myself that any of the colourless corpuscles made their way out of the interior of the veins or capillaries. I have subjected the mesentery to examination for the purpose, both when it has been drawn out for the first time and disposed under the microscope, and again a day, two days, and three days after,* and I have seen colourless corpuscles accumulated

---

* After each sitting of a few hours' duration the intestine and mesentery were returned into the abdomen, and the opening into this cavity closed by suture. The frog was then put into a wide-mouthed bottle and kept there until the next examination, when it was again secured under the microscope, and the intestine with the mesentery prolapsed as before.

in great numbers on the walls of the dilated veins, as I
have also seen in the web. No doubt the accumulated
colourless corpuscles appear at many points in the
course of the veins as if accumulated outside the vessel,
but I could always recognise the wall outside by
adjusting the focus of the microscope on it. And when,
at a later stage, I have not been able to recognise the
outline of the wall outside the accumulated colourless
corpuscles, I was as little successful in recognising an
outline of the wall inside, for the wall had by that
time given way, as was the case under the circumstances described in the web of the frog, in the quotation above made from my Essay in Guy's Hospital
Reports. No doubt also, some of the corpuscles are
seen at every instant to become detached from the rest
that are stagnant, and, when the outline of the vessel
is not in focus, seem as if escaping from the interior;
but that they have not so escaped is immediately
observed by bringing the wall into focus. No doubt,
also, small diverticula of the wall of the vein may be
noticed in which a colourless corpuscle is sometimes
seen to become lodged; but such I have observed at
last to be carried away by the stream of the blood;
sometimes I have seen a second colourless corpuscle
strike it and adhere to it, whereupon both were carried
away together.

The amœba-like changes of form and the movements
which the colourless corpuscles present, and which Dr.
Cohnheim views as characteristic of an essential stage
in the inflammatory and suppurative processes, appear

to be similar in their nature to those described in the extracts from my papers in the Philosophical Transactions above given, as occurring after the blood has been drawn from the body. The contractions and out-shootings are due to an inherent endowment of the cell-wall, but that endowment is called into unusual activity by the changed condition of the blood. In blood drawn from the body, the changes of shape which the colourless corpuscles undergo of course lead to nothing, but in the case of blood effused into the interstices of the tissues some purpose is no doubt served by them. According to Dr. Cohnheim, they play an important part in the alleged escape of the colourless corpuscles from the interior of veins and capillaries just mentioned.

He says that amœba-like changes of shape do not present themselves as long as the colourless corpuscles are flowing in the stream, but only after they have been stagnant on the walls of the veins and capillaries for some time, and that they are rarely seen in the web, as stagnation there does not continue long enough.

When, then, stagnation has taken place, the hitherto globular colourless corpuscles begin to undergo their changes of form and to pierce through the walls of the vessel in the manner above described.

Within the vessels of the frog's web, I have observed colourless corpuscles continue stagnant for a much longer time than it is possible to observe them in the mesentery, in consequence of the death of the frog supervening in a few days on the injury inflicted on it by opening the abdomen and drawing out the intestine. Notwithstanding the length of time I have had under view the colourless corpuscles stagnant within the veins of the web, I must say that I never noticed them undergo any changes of form other than what I have described at pp. 127-128. As little have I seen any other changes of form within the small veins of the mesentery.* But supposing such changes do occur, at least in the case of colourless corpuscles stagnant within the veins of the mesentery, it cannot be admitted that their cell-wall shoots out its amœba-like processes with any force such as is calculated to penetrate the wall of the vessel in the manner Dr. Cohnheim seems to think.

Outside the vessels of the mesentery I have seen, as Dr. Cohnheim describes, granule cells or colourless corpuscles, exhibiting amœba-like changes of shape and position, but as they appeared in company with red corpuscles, I considered them as the colourless corpuscles of the extravasated blood. Those which I saw on the mesentery at a later period of the case were, no doubt, new formations. Indeed, Dr. Cohnheim's

---

* My observations on the state of the blood-vessels generally in the mesentery were made chiefly with an objective, ⅓rd of an inch, by the Messrs. Merz, of Munich ; but I have watched the colourless corpuscles in small veins under objectives ¼ and ⅛th of an inch, by Dallmeyer, with a high power of eye-piece as well as with the lowest. I have thus viewed the objects magnified as much as from 500 to 900 diameters. Here I cannot refrain from bearing testimony to the excellence of Mr. Dallmeyer's objectives. The eighth of an inch which he made for me is an exquisite instrument. Being provided with a stop, it admits of being advantageously used in such examinations as those under notice.

own description* of the first appearance of the corpuscles, from the boss-like protuberances on the outside
of the vessels is very suggestive of this view.

If colourless corpuscles do make their way out
through the walls of a vein or capillary in any number,
the phenomenon could scarcely have escaped my
notice. But admitting that they do, it surely cannot
be alleged that the walls of the vessel remain entire.

The corpuscles, like colourless blood corpuscles, which
composed the newly formed white structure with which
the wounds of the web, described at pp. 147—9, were
filled up, were certainly not colourless corpuscles which
had escaped from the blood in the interior of vessels,
but were newly formed cells. Neither were the cells
contained in the matter described in the following
extract from my Essay in Guy's Hospital Reports
(p. 44), escaped colourless corpuscles :—

"In inflammation from the action of a strong solution of salt on the web, exudation is manifested by
opacity and thickening, and, in a greater degree, by a
collection of fluid between the two layers of skin. In a
web to which a strong solution of salt had been applied
three days before, and in which congestion and stagnation had been produced, the circulation was found
re-established in great part, except at the free margin
of the two largest webs, where there were collections of
fluid between the layers of skin, giving rise to the
appearance of blisters. The fluid contained in these
blisters having been evacuated and placed under the

* Virchow's Archiv, ut supra, pp. 33-9.

microscope, was found to present elementary granules, nuclei surrounded by an amorphous substance, and completely formed nucleated cells. The latter were very pale, finely granular, and about $\frac{1}{1800}$th or $\frac{1}{16000}$th of an inch in diameter."

Through the walls of capillaries red corpuscles, according to Dr. Cohnheim, pass out as well as colourless ones ; but this, he says, does not take place until after the colourless have escaped.

I have often seen a few red corpuscles pass one after another from a capillary in which the blood was flowing into an empty anastomosing capillary previously nearly invisible. If colourless corpuscles, however, do escape through pores in unruptured walls of capillaries, there can be no difficulty in admitting that red corpuscles do so likewise, especially if altered in any such manner as red corpuscles are by the action of pus, as above described, p. 286.*

Dr. Cohnheim has never seen corpuscles, either colourless or red, pass out through the walls of arteries.

---

* My colleague, Dr. Bastian, assures me that he has observed the escape of blood corpuscles, both red and colourless, in great numbers through the walls of veins and capillaries in the mesentery of frogs under the influence of curare, as described by Dr. Cohnheim. That such an occurrence depends on some change in the condition of the blood, or the blood-vessels, or of both, induced by the action of that poison, and perhaps also by the drying action of the air, and by the pressure to which the mesentery is subjected by being kept stretched under the microscope for a day or two, appears to me probable. I have not, however, examined into the point, my present observations having been confined to the phenomena of inflammation as they present themselves under conditions as normal as possible.

KERATITIS OR INFLAMMATION OF THE CORNEA ACCORDING TO DR. COHNHEIM'S VIEW.

Dr. Cohnheim recognises in keratitis the attending vascularity in the adjacent part of the white of the eye, and admits that the matter exuded into the interstices of the cornea comes from the blood in the congested vessels. But, in accordance with his peculiar views, the exuded matter consists of colourless corpuscles of the blood which have escaped bodily through the walls of the congested vessels and made their way, by means of their amœba-like movements, into the interstices of the cornea.

Dr. Cohnheim maintains that there is nothing to warrant the opinion entertained by Virchow that the pus corpuscles in an abscess of the cornea are developed by a process of division from the cells pre-existing in the interstices of that structure.

While thus repudiating Professor Virchow's view of inflammation, so remarkable for its extremeness in one direction, Dr. Cohnheim, it appears to me, promulgates a view of the subject as remarkable for its extremeness in the opposite direction.

INDEX.

THE END.

www.ingramcontent.com/pod-product-compliance
Lightning Source LLC
Chambersburg PA
CBHW021501210326
41599CB00012B/1089